# The Human Embryonic Stem Cell Debate

**Basic Bioethics**
Glenn McGee and Arthur Caplan, editors

*Pricing Life*
Peter A. Ubel

*Bioethics: Ancient Themes in Contemporary Issues*
edited by Mark G. Kuczewski and Ronald Polansky

*The Human Embryonic Stem Cell Debate: Science, Ethics, and Public Policy*
edited by Suzanne Holland, Karen Lebacqz, and Laurie Zoloth

# The Human Embryonic Stem Cell Debate
Science, Ethics, and Public Policy

Edited by Suzanne Holland, Karen Lebacqz,
and Laurie Zoloth

A Bradford Book
The MIT Press
Cambridge, Massachusetts
London, England

This book was set in Adobe Sabon in QuarkXPress by Asco Typesetters, Hong Kong and was printed and bound in the United States of America.

Library of Congress Cataloging-in-Publication Data

The human embryonic stem cell debate : science, ethics, and public policy / edited by Suzanne Holland, Karen Lebacqz, and Laurie Zoloth.
    p.   cm.
  Includes index.
  ISBN 0-262-08299-3 (hc.)—ISBN 0-262-58208-2 (pbk.)
  1. Human embryo—Research—Moral and ethical aspects. 2. Stem cells—Research—Moral and ethical aspects. I. Holland, Suzanne. II. Lebacqz, Karen, 1945- III. Zoloth, Laurie.
QM608 .H86   2001
174′.2—dc21                                              2001030624

I have learned much from my teachers
More from my friends
And from my students
Most of all.

Rabbi Hanina
*Babylonian Talmud*, Ta'anith 7a

# Contents

# Series Foreword

We are pleased to present the third volume in the series Basic Bioethics. The series presents innovative books in bioethics to a broad audience and introduces seminal scholarly manuscripts, state-of-the-art reference works, and textbooks. Such broad areas as the philosophy of medicine, advancing genetics and biotechnology, end of life care, health and social policy, and the empirical study of biomedical life will be engaged.

Glenn McGee
Arthur Caplan

*Basic Bioethics Series Editorial Board*
Tod S. Chambers
Carl Elliot
Susan Dorr Goold
Mark Kuczewski
Herman Saatkamp

# Acknowledgments

The editors could not have compiled this volume, reflecting the first years of debate on stem cell research, without the dedication and cooperation of a number of extraordinary people. First and foremost, we wish to thank Audrey Englert, secretary to the faculty of Pacific School of Religion, who worked tirelessly to format, rescue, and copy edit hundreds of pages under severe time constraints, putting in untold overtime hours. Second, we are indebted to the authors, whose generosity and intensity of scholarship allowed us to create what we intend to be a first critical contribution to the public discourse so integral to our society, to medicine, and to the field of bioethics. We also thank our home institutions, the University of Puget Sound, Pacific School of Religion, and San Francisco State University, for their cooperation and support—and for the sabbaticals that enabled us to devote extensive time to this effort. Our families, children, and friends were invaluable in understanding the excitement and urgency with which we pursued this work. Carolyn Gray Anderson of MIT Press had the vision to see and support the timeliness of this work. Finally, we wish to thank those behind the scenes—the researchers and patients hoping for new therapies, who shared their stories and sometimes their laboratories with us. Our venture into this historic dilemma is a leap of faith that could not have been accomplished without the integrity and willing help of all these colleagues, and to them we are deeply grateful.

# Introduction

This book captures some of the first foundational work surrounding the new and controversial discovery of human pluripotent stem cells. It is also an invitation to the reader to join a continuing dialogue that has broadened considerably and sparked strong public reaction.

When derivation of human pluripotent stem cells was announced in November 1998 it caught most of the scientific community and the public by surprise. Although work in animals had successfully isolated stem cells in a number of species, the search for human pluripotent stem cells seemed elusive at best. The simultaneous announcement of the isolation of human embryonic stem (hES) cells (Thomson et al. 1998) and human embryonic germ (hEG) cells (Shamblott et al. 1998) set off a storm of controversy. Forces quickly mobilized: President Clinton asked the National Bioethics Advisory Commission (NBAC) to undertake a thorough review of issues associated with stem cell research (NBAC 1999), religious leaders reiterated their opposition to creation of embryos for research or to destruction of embryos in research, and ethicists began to study the issues involved.

## A Short History

We put this book together for several reasons, not the least of which is that, in one way or another, we have been engaged in the ethical debate surrounding this scientific breakthrough. In late 1996, representatives of a small private biotechology company in Menlo Park, Geron Corporation, came to the Graduate Theological Union in Berkeley, California, seeking conversation about the ethical dimensions of work they were about to undertake. All three editors of this volume were part of those early conversations. Two of us (Lebacqz and Zoloth) subsequently became members of the Geron Ethics Advisory Board (GEAB). As early participants in an intriguing and morally complex dialogue, we were sometimes apprised of scientific achievements before those achievements were widely known. Because research of both Thomson's and Gearhart's labs had been sponsored by Geron, for example, the GEAB knew of the forthcoming announcement of the derivation of human pluripotent stem cells before this discovery was common knowledge. We first learned about the hES cell and hEG cell research in August 1998, and we found ourselves hard pressed to sort through the ethical issues at stake in the scant few weeks before the public announcement. By September we had determined that all of us could support the research, and in October 1998 we adopted a set of minimal guidelines for the work. Those guidelines and an accompanying rationale were published in early 1999 along with some commentaries (GEAB 1999) We waited with anticipation for the NBAC, the American Academy for the Advancement of Science (AAAS), and other bodies with broader representation to enter the public debate and to revise, refine, or reject our work.

During spring of 1999, NBAC held public hearings on stem cell research and drafted its preliminary report, which was published in September of that year. Several members of GEAB were among those who testified before NBAC, as were a number of other contributors to this volume. Because NBAC stands as the most important open public forum for discussion of ethical issues surrounding this work, several chapters are either drawn from the testimony or take NBAC's recommendations as the basic public policy that must be addressed and critiqued.

At the same time that NBAC was conducting its deliberations, the AAAS held a series of meetings in which ethical and scientific issues around stem cells were discussed and policy recommendations issued (Chapman et al. 1999). Zoloth and Lebacqz were involved in those meetings. In addition, the National Institutes of Health (NIH) conducted its own investigation into the legalities of permitting the use of federal funds for stem cell research (NIH 1999a) and began drafting guidelines for human pluripotent stem cell research (U.S. Department of Health and Human Services [DHHS] 1999) that were made final in August 2000

(DHHS 2000). The immediate attention of so many prominent public bodies indicates the significance of this research.

### What Are hES and hEG Cells and Why Are They so Important?

Stem cells have the capacity for prolonged self-renewal and can produce at least one type of highly differentiated or specialized descendant (Watt and Hogan 2000; Weissman 2000). In adults, they are present in many tissues, blood and skin, for example. They enable the body to regenerate tissues or cells such as bone marrow. Until recently, it was commonly assumed that stem cells from specific tissues could generate only tissues of those types; hence, they were understood to be powerful in capability, but limited in direction. Recent reports suggest that stem cells from adult tissue may be more malleable than previously thought, however; for example, those from adult mouse brains are able to generate other than brain tissue (Clarke et al. 2000). Of course, it is by no means conclusive that what is true of mice stem cells will hold true for humans; experience with cloning technology suggests how difficult the transition between species can be.

Cells of the early embryo are not limited in the way that adult stem cells were assumed to be. As the fertilized egg divides, each cell is able to be separated out and form an entire new organism. Hence these cells (or blastomeres) are *totipotent*—they have potential to form any and all human tissues and to become a complete organism (NIH 1999b; Gage 2000). At the point at which dividing cells develop into a hollow ball, the embryo is called a blastocyst. The hES cells are derived by destroying the outer shell of the blastocyst, which would normally become the placenta, and culturing cells from the inner cell mass (see Thomson's chapter). A second source of cells with similar potential is the gonadal ridge of the aborted fetus (Shamblott et al. 1998). These cells would have developed into germ cells; hence the designation, hEG. Although both types of cells are discussed in this volume, it is not yet clear that hEG cells will have the same capacities and characteristics that make hES cells so important. Hence, our primary focus is on hES cell research.

The hES cells are important because they have certain critical characteristics. Most important they are *pluripotent*—they are able to develop into many types of tissues (thus, they are also sometimes called pluripotent stem cells, or PSCs). They are also *immortal*—able to continue dividing indefinitely without losing their genetic structure. Moreover, hES cells are *malleable*—they can be manipulated without losing cell function. Indeed, studies with animal stem cells suggest that they can be moved into another blastocyst and it will continue its development. Finally, they express the enzyme telomerase, which allows cells to grow and divide.

All of these qualities make the study of these cells extremely important for medical science. Put simply, because hES cells appear to be able to become any kind of tissue, once mechanisms for differentiation are understood, they might provide banks of skin, bone, liver, and other tissues to repair or replace body parts (Hall 2000). Equally important is the possibility of creating cultures of tissue for testing new drugs.

Given these potential uses, major bodies such as the NBAC, the AAAS and the NIH acknowledged hES cell research as one of the most significant breakthroughs of the century and as holding promise of a new era of medicine. Indeed, supporters and detractors alike concur in seeing the potential importance of these cells. Hailed as the "breakthrough of the year" by *Science* (Vogel 1999), several commentators suggest that hES research will usher in a new era of "regenerative medicine." Further discussion of this topic can be found in chapters by Thomson and Okarma.

**Why Is hES Research Controversial?**

In spite of this importance—and perhaps because of it—hES cell research has proved to be one of the most controversial developments of the last decades. Controversy arises for many reasons. The research touches deep questions about the nature of human life, limits of interventions into human cells and tissues, and the meaning of our corporate existence. We have probably only begun to identify and approach the important ethical issues. However, the immediate controversy revolved around a cluster of difficult and sometimes seemingly intractable questions.

First, the derivation of stem cells from living embryos that are destroyed in the process or from fetuses that have been aborted touched off a firestorm of controversy because it tapped into the contentious debate in this country around abortion. For many people of faith, an embryo is fully a human being and therefore may not be killed. Furthermore, any activity that appears to support or contribute to abortion is anathema to many people, for religious or other reasons. Sensitivity to such concerns and their political implications led to adoption of a federal policy that no public monies may be expended to support research that involves destruction of an embryo. Thus, the first contentious issue concerned the status of embryos and the possibility of their destruction. What is the status of an embryo and should it be protected against destruction? This question has dominated the ethical discussion and a number of chapters address it (Farley, Meilaender, Lebacqz, Young, Shannon).

A related question is whether there should be limits to what humans may do to themselves and to human genes, human bodies, and developing human organisms. Here, the focus is not so much on the status of the embryo as on the perennial question of pride and humility. Should certain limits be honored lest we overstep the bounds of what it means to be human? Many people feel uneasiness when some lines are crossed, for example, when life stretches well past a "normal" span, when postmenopausal women are given the possibility of bearing children, when fertility drugs result in births of seven or eight children at once. Whereas our society generally welcomes medical advances, concern is growing that we may be breaching creaturely limits by our ever-expanding medical technologies. Possible links between stem cell research and somatic cell nuclear transfer or cloning technology raise grave concerns for some (see Parens). Broad issues of human restraint and human ingenuity create ethical conflict in this arena.

Because of federal policy prohibiting use of public monies for this research, to date the development of hES cells has depended on private funding. At the time of this writing, however, this issue remains somewhat open-ended. In August, 2000, the NIH published its Guidelines for Research Using Human Pluripotent Stem Cells in the Federal Register (NIH 2000). The new regulations make it possible for government to fund research on human pluripotent stem cell lines derived from embryos or fetal tissue. The distinction is one of deriving rather than creating such cell lines, regulating use of public funds by degrees. In March of this year, the NIH began receiving the first round of grant applications under these new guidelines, although it remains to be seen whether President George W. Bush will take measures to block these NIH regulations. Should he do so, funding for hES cell research will remain exclusively in the domain of the private sector. Private funding for research raises other ethical issues. Are oversight and attention to ethical concerns sufficient when research is carried on in the secrecy of the private arena rather than under the scrutiny of government funding agencies such as NIH? Whereas much controversy focuses on the derivation of stem cells from embryos and fetuses, some are concerned about lack of public oversight of this research. This concern was heightened when another biotechnology company, Advanced Cell Technology, claimed that it had created a cow-human hybrid. It was partly this announcement that led President Clinton to request that NBAC study issues around stem cell research (NBAC 1999). Creation of such a hybrid would be anathema to many people, and raises issues about violation of ethical standards where there is little public oversight of research. Although both Thomson and Gearhart had to submit research proposals to several layers of institutional review boards at their respective universities, Geron took the unusual step of establishing the GEAB to provide another layer of review. Corporations and institutions that do not receive federal funds for research are not mandated by law to have such review boards, however. Thus, another large issue is how to ensure appropriate reviews and whether the federal ban on research involving creation or destruction of embryos actually undermines good review. Several chapters address the appropriate layers of review (Cohen, McLean, Wolpe and McGee).

Another cluster of issues centers on context and consent for research. Particularly in the absence of public oversight, it becomes important to know whether private mechanisms for protection were invoked. Was consent obtained? Was the consent process adequate? Did women or couples who donated 'excess' embryos from in vitro fertilization really understand what would be done with the embryos in research? Did women giving consent for use of aborted fetuses to obtain hEG cells understand what was involved? Who should give consent for the use of embryos that have been determined to be excess and not needed in in vitro fertilization—the woman donating eggs? the woman receiving them for implantation? their partners? What constitutes "excess" embryos?

Who should give consent for the use of tissue from aborted fetuses, and under what conditions? Thus, the role of women or couples from whom the fetuses or embryos are obtained, and uncritical acceptance of in vitro practices that underlie access to embryos are major concerns for some philosophers (see Baylis).

Yet another issue concerns the status of the newly created hES cell. As Thomson makes clear in his chapter, this cell is not precisely the same as cells existing in the blastocyst. What, then, are its character and status? Or "What's in the Dish?" (McGee and Caplan 1999). Is the hES cell more akin to an embryo or to an ordinary somatic cell? Of what significance is its pluripotency? What status should it have, and what protections should be brought to bear? Is it something that should be able to be patented? Such questions trouble some observers.

Equally problematic is how all these issues should be decided. Who should determine the status of a newly created cell? In a multicultural society with many religious traditions represented, how should conflicting views of the value and status of embryos and stem cells be adjudicated? In a pluralistic society based on liberal philosophy, such as the United States, governments are not supposed to intervene in private decisions. Rather than imposing specific views or values, governments content themselves with setting up procedures that are intended to ensure fairness. But when something is as deeply central to human life as stem cells may prove to be, retreat into procedural justice is inadequate in the views of many. It is not easy to allow value decisions to remain in the private arena when that arena has such impact on the common good. The very procedures for decision making become contentious (Fletcher).

Finally, difficult issues relate to what language to use to describe what we are doing both in science and in ethics. Scientists speaking of stem cells tend to use words such as pluripotent and totipotent. The meaning of such terms is not always agreed or clear. For example, totipotent is typically reserved for the concept that a cell could become an entire living organism; however, it has been used to indicate the capability to produce tissue of all types, which makes its meaning closer to pluripotency. Even if scientific meaning can be agreed, adopting scientific terminology carries its own risks and dangers. Scientific language will tend to dominate the discussion, but as some commentators have shown, adoption of any language as normative tends to shut out other considerations that can be carried only by alternative language (Evans 1999). What it means to speak of the self, for example, may differ in science and in religious discourse. Do we even have a language adequate to describe what we might do with this new technology? For example, ethicists and scientists alike tend to distinguish gene therapy from genetic enhancement (Walters and Palmer 1997). But this distinction is already tenuous and may collapse entirely if we develop the capacity to reprogram cells and change the underlying genetic structure of our tissues.

These are only some of the many ethical issues that are raised by contributors to this volume. Other equally important concerns arise: what to do about the inevitability of errors and how to evaluate our readiness to use a technology in light of that inevitability; unintended consequences and whether, ethically speaking, their likelihood should prevent our taking certain steps; profit and commodification of important parts of human living; and the perennial and intractable issue of social justice and how to distribute the benefits and burdens that will inevitably be associated with this research. Several authors address social justice (Fletcher, McLean, Zoloth), but considerably more work remains to be done on this and other topics that to date have received less attention than the status of the embryo (Holland).

## The Book

Given all of these concerns and issues, it is no wonder that Regalado (1998) called the search for hES cells "the most intriguing, controversial, underfunded and hush-hush of scientific pursuits." This book will neither stem the controversy nor solve the funding issue, but it may contribute to lessening the intrigue and ending the hush-hush nature of the pursuit.

Our design is simple. In part I we offer three essays to ground the ethical discussion. Thomas B. Okarma, president and CEO of Geron Corporation, puts hES and hEG cell research into context by looking at how it supports larger scientific and industry goals directed toward health and healing. James A. Thomson, whose team first isolated and cultured hES cells, details the science of the cells and offers a vision from the scientist's perspective as to why their discovery is so important. Finally,

John C. Fletcher, long known for his work in ethical aspects of advances in genetics, puts the controversy into historical context by comparing contemporary debate with work of the National Commission for the Protection of Human Subjects of Biomedical and Behavioral Research in the mid-1970s.

Part II offers an introduction to ethical issues in hES cell research. Erik Parens's chapter, originally prepared as background for NBAC, was instrumental in NBAC's own reasoning and analysis. He presents a careful and critical analysis of the implications of stem cell research. He notes in particular the links between this research and germline intervention, somatic cell nuclear transfer, and other new technologies. Writing from a Canadian context, Françoise Baylis argues that NBAC relied on a problematic understanding of "respect," an uncritical acceptance of fertility clinic practice, and a questionable decision-making process. In a second contribution to this volume, John Fletcher holds that NBAC failed to present evidence to back up claims for the promise of the technology, that its use of abortion as an analogy is flawed, and that serious justice questions remain to be addressed. Suzanne Holland zeroes in on one of those justice questions—the status of women and people of color. She charges that the debate on hES cell research neglects issues crucial from a feminist perspective. In particular, she shows how the acceptance of a public-private split not only undermines good public policy but also works to the disadvantage of the marginalized. Taken together, these chapters uncover a number of ethical dilemmas and suggest modes of analysis that are important in the field of ethics—philosophical reasoning, historical comparison, and feminist analysis.

Authors in part III offer angles of vision both on ethical questions and modes of analysis. The introduction of religious perspectives suggests new questions and sometimes new ways of thinking about ethical analysis. Eliot N. Dorff's testimony before NBAC, reprinted with slight modifications, stressed fundamental Jewish assumptions such as the duty to heal and the need for humility, and suggests that there may even be a duty to proceed with the research. Offering another Jewish perspective is Laurie Zoloth, who demonstrates how Jewish halachic reasoning works by offering explicit textual analysis regarding the status of the fetus, obligations to the dying, and the mandate to heal.

Margaret A. Farley's testimony to NBAC noted the variety of possible Roman Catholic responses. For example, the status of the fetus becomes critically important. A second Roman Catholic view is provided by Michael M. Mendiola, who identifies resources within that faith tradition that provide tentativeness in ethical reasoning and an ethic of toleration. Both these thinkers suggest modes of reasoning available within their tradition that offer a wider range of possible responses than is often attributed to Catholic thought on bioethical issues that touch on the status of the embryo.

One of the central ethical questions regards respect for embryonic life. Several chapters take on the issue directly. Ted Peters holds that both sides of the debate depend implicitly on a notion of a significant difference between pluripotent and totipotent cells. Assuming that this difference will be undermined eventually by scientific developments, he suggests that the dignity and value of the hES or any other cell cannot be linked to its potential. A Christian perspective, he concludes, offers an alternative understanding. Gilbert Meilaender's Protestant testimony before NBAC cautioned that we must speak truthfully about what we are doing. Urging that the embryo is among the weak in society, it is not possible to respect the embryo or fetus while intending its destruction. Both Karen Lebacqz and Ernlé W. D. Young develop a specific analysis of respect. Drawing on practices and using casuistic reasoning, Lebacqz suggests that respect for the embryo is not obviated by the research in question, even though that research involves destruction of the embryo. Drawing on secular philosophical literature, Young elaborates and applies Mary Ann Warren's principles of respect to provide a secular view on rights, respect, and stem cell research.

Most of these authors find room for permissible stem cell research, but their pathways to such a conclusion are so different as to suggest many unresolved issues in the debate, and that further conversation is mandated. Both modes of ethical discourse and specific policy recommendations and ethical judgments vary widely among these views. All of them take the status of the embryo as critically important, but some hold that status in tension with other values such as the mandate to heal, whereas others maintain that a theological view mandates particular commitment to the weak and vulnerable or to new understanding of the origins of respect. Others, finally, believe that respect must be nuanced according to the nature of what is being respected.

In part IV, commentators take up another set of questions: public policy issues and questions of review, oversight, and social justice. Thomas A. Shannon contends that micro issues such as the status of the fetus might be resolved, but that important macro issues remain, specifically issues of justice. Paul Root Wolpe and Glenn McGee suggest that the debate has been framed by experts and that it is imperative that it be extended into grass roots levels. Margaret R. McLean argues that the nature of this work is such that local overview is not sufficient. She offers five basic components necessary for fair public policy around this important technology. Similarly, according to Cynthia B. Cohen, only a special review structure at the national level will do. Laurie Zoloth urges new vision and moral imagination if we are to keep up with the technology at all, and states that we should be willing to slow down technology in the interests of cultivating a vision adequate to bring justice.

These commentators do not all agree on what kinds of review or public policies would be most fair. What they do agree on is the need for attention to questions of social justice and for the provision of structures of oversight adequate to ensure more justice than a haphazard system will bring.

Opening this book is therefore entering into a room in which a new conversation is taking place. It is, in large part, a conversation about the ethical, scientific, philosophic, and religious meaning of who we are as human beings and what our fate will be in the new century. Whereas the conversation centers on one part of that fate—the science of genetic or regenerative medicine—the implications of this science are much broader. At stake are issues regarding who should make important decisions about approving controversial research, whether it is possible to continue demarcating different types of cells in the way that we have, what constitutes human dignity and respect, what justice requires in the new millennium, and how ethical reasoning ought to proceed. This book captures only part of the important conversation. Some of the foundational work is here, and we hope it is the beginning of a broader public dialogue and a careful assessment of the promises and pitfalls of the new technology.

## References

Chapman, A. R., Frankel, M. S., and Garfinkle, M. S. 1999. *Stem Cell Research and Applications: Monitoring the Frontiers of Biomedical Research.* Washington, DC: American Association for the Advancement of Science and the Institute for Civil Society.

Clarke, D. L., 2000. Generalized potential of adult neural stem cells. *Science* 288(5471): 1660–1663.

Evans, J. H. 1999. The uneven playing field of the dialogue on patenting. In *Perspectives on Genetic Patenting.* A. Chapman, ed. Washington, DC: American Association for the Advancement of Science, pp. 57–73.

Gage, F. H. 2000. Mammalian neural stem cells. *Science* 287(5457): 1433–1438.

Geron Ethics Advisory Board. 1999. Research with human embryonic stem cells: Ethical considerations. *Hastings Center Report* 29(2): 31–36.

Hall, S. S. 2000. The recycled generation. *New York Times Magazine,* January 30.

McGee, G. and Caplan, A. L. 1999. What's in the Dish? *Hastings Center Report* 29(2): 36–38.

National Bioethics Advisory Commission. 1999. *Ethical Issues in Human Stem Cell Research.* Vol. 1. *Report and Recommendations.* Rockville, MD: National Bioethics Advisory Commission.

National Institutes of Health. 1999a. NIH backgrounder: Fact sheet on human pluripotent stem cell research guidelines. www.nih.gov/news/stemcell/factsheet. htm.

National Institutes of Health. 1999b. Stem cells: A primer. www.nih.gov/news/ stemcell/primer.htm.

National Institutes of Health. 2000. Guidelines for Research Using Human Pluripotent Stem Cells. *Federal Register*, August 25. (www.nih.gov/news/stemcell/ stemcellguidelines.html).

Regaldo, A. 1998. The troubled hunt for the ultimate cell. *Technology Review* (July/August).

Shamblott, M. J. Axelman, J., Wang, S. Bugg, E. M., Littlefield, J. W., Donovan, P. J., Blumenthal, P. D., Huggins, G. R., and Gearhart, J. 1998. Derivation of pluripotent stem cells from cultured human primordial germ cells. *Proceedings of the National Academy of Science (USA)* 95: 13726–13731.

Thomson, J. A., Liskovitz-Eldor, J., Shapiro, S. S., Waknitz, M. A., Swiergiel, J. J., Marshall, V. S., and Jones, J. J. 1998. Embryonic stem cell lines derived from human blastocysts. *Science* 282: 1145–1147.

U.S. Department of Health and Human Services. 1999. Draft National Institutes of Health guidelines for research involving human pluripotent stem cells. *Federal Register* 64(231): 67576–67579.

U.S. Department of Health and Human Services. 2000. *Stem Cell Guidelines.* www.nih.gov/news/stemcell/stemcellguidelines.

Vogel, G. 1999. Breakthrough of the year: Capturing the promise of youth. *Science* 286(5448): 2238–2239.

Walters, L. and Palmer, J. G. 1997. *The Ethics of Human Gene Therapy.* New York: Oxford University Press.

Watt, F. M. and Hogan, B. L. M. 2000. Out of Eden: Stem cells and their niches. *Science* 287(5457): 1427–1430.

Weissman, I. L. 2000. Translating stem and progenitor cell biology to the clinic: Barriers and opportunities. *Science* 287(5457): 1442–1446.

# I

## The Science and Background of Human Embryonic Stem Cells

# 1

## Human Embryonic Stem Cells: A Primer on the Technology and Its Medical Applications

Thomas B. Okarma

### The Breakthrough and Its Potential

The new millennium has brought with it extraordinary advances in biomedical sciences. Completion of the human genome sequence (as well as genome sequences of several lower species), microarray technology to measure simultaneously the expression of thousands of genes in single experiments, and improved efficiencies of drug discovery aided by rapid-parallel compound synthesis and ultrahigh-throughput screening technologies are all helping to bring us significantly closer to realizing totally new therapeutic approaches for major chronic diseases.

Among these outstanding advancements is successful derivation of the human embryonic stem (hES) cell (Thomson 1998, 282), a self-renewing cell line that gives rise to all cells and tissues of the body. The potential for these cells is to allow permanent repair of failing organs by injecting healthy functional cells developed from them, an approach called regenerative medicine. The significance would be to broaden the definition of medical therapy from simply halting the progression of acute or chronic disease to include restoration of lost organ function. To illustrate, patients who suffered a myocardial infarction would be discharged from hospital not only with immediate progression of the infarct stopped, but also with a repaired heart, the function of which would be restored to pre-infarct state. Patients with stroke or spinal cord injuries could receive cell-based treatments that would restore central nervous system function, thereby enabling them to maintain functional independence. Regenerative medicine would be a totally new value paradigm for clinical therapeutics.

Restoration of lost organ function cannot be achieved through traditional drug therapies. Damage to heart tissue, brain tissue, or other organs caused by hemorrhage, blood clots, or other damaging processes is usually so extensive as to be beyond the reach of drugs. The usual mechanism of action of most drugs is to alter aspects of a cell's metabolism, not to cause growth of healthy replacement cells that restore function. In the case of catastrophic diseases, cellular substrate of the organs themselves is irreversibly damaged and replaced with dysfunctional scar tissue, leaving the organ severely compromised. Rudolph Virchow stated in the 1850s that "all cells come from cells" (McLaren 2000, 288). This principle remains true today. The only way to restore cellular function in an organ is literally to replace the lost cells.

## The Biology Behind the Breakthrough

Certain organ systems of the human body are capable of regeneration throughout life. We are constantly shedding and regrowing new layers of skin. Blood cells die and are replaced with new ones originating from bone marrow. Cells that line the gastrointestinal tract are removed and replaced by new ones. Menstruating women replace the cellular lining of their uterus each month. What is the source of these new cells that repopulate the skin, blood, intestines, and uterus? Where do these source cells come from and what are the mechanisms by which their growth and differentiation are regulated?

The answers to these questions take us to the topic of stem cells, defined as cells that can both renew themselves in the undifferentiated state as well as differentiate into descendent cells that have a specific function. The prevailing view is that among organs with self-renewal capability, resident stem cells are capable of periodically (or continuously) providing new populations of functional, differentiated cells that can replace those lost by normal physiologic turnover or even some types of catastrophic losses due to injury or disease. When we donate blood, hematopoietic feedback loops respond by stimulating bone marrow stem cells to accelerate production of replacement blood cells. This concept was developed and applied therapeutically in allogeneic bone marrow transplantation. Here, hematopoietic stem cells from one healthy individual are transferred to the patient, thereby repopulating the patient's entire blood system.

Most stem cells have limited potential to form only certain differentiated progeny cells. Hematopoietic stem cells can produce only blood cells, skin stem cells can produce only skin cells, and so on. This principle generally holds despite provocative observations that, under certain circumstances, a blood stem cell, for example, can be coaxed to produce a nerve cell (Bjornson et al. 1999, 283) or liver cell (Petersen et al. 1999, 284). Restriction of differentiation potential, however, is characteristic of most stem cells that have been isolated and studied to date. The only certain exception is the embryonic stem cell, which can give rise to literally all cells and tissues of the body. Embryonic stem cells are therefore called pluripotent. Although derived from very early embryos, they are not themselves embryos and cannot under any circumstance develop by themselves into whole animals or humans. They are therefore not totipotent, as is the zygote (fertilized egg), which is formed at the time of conception and on its own forms an embryo and placenta in the uterus.

The second important fundamental characteristic of stem cells is self-renewal—their ability to divide asynchronously into one differentiated daughter cell and one stem cell-like daughter cell. Herein lies the second distinction between hES cells and all other stem cells discovered to date. Hematopoietic stem cells, for example, can be removed from bone marrow or blood and cultured in the laboratory. Under these conditions, however, the cells eventually cease dividing and no longer self-renew. In contrast, hES cells have been grown continuously in laboratory conditions for over two years without losing their ability to self-renew or to form all cells and tissues of the body (Amit et al. 2000).

Because of pluripotency and infinite self-renewal, hES cells are perhaps the most extraordinary cells ever discovered. Their discovery certainly qualifies as one of the major breakthroughs in biomedicine.

## Implications for Biomedicine

### Understanding Human Developmental Biology
For obvious ethical and practical reasons, it is not possible to study rigorously the molecular biology of human embryonic development. However, hES cells can be studied to define the genetic blueprint used by

nature to build a human body, cell by cell and tissue by tissue. These studies will increase our understanding of the molecular mechanisms of normal development that will provide a foundation for understanding fetal developmental abnormalities. What goes wrong with this natural genetic blueprint to result in early miscarriage or the birth of children with congenital abnormalities? We have learned much from studying the development of mouse and other animal embryos, but they are at best an approximation to the human. Despite the surprising degree of genetic similarity between humans and laboratory animals, human developmental biology is unique.

Advancing our fundamental understanding of human reproductive and developmental biology is an important U.S. health care objective. Fertility disorders affect one of every six couples in the United States trying to conceive. Premature pregnancy loss is estimated to occur in up to 15 percent of recognized pregnancies in the United States, and birth defects afflict 3 percent of live births in this country. Until now, early developmental events that occur naturally during embryogenesis have been inaccessible to direct study. The availability of hES cells will facilitate molecular understanding of how specific tissues and organs develop without conducting direct research on human embryos or fetuses. Furthermore, genes that fundamentally control tissue differentiation may be identified by applying genomic technologies to cultured hES cells as they differentiate and grow into a variety of cell types. Identification of genes that control normal tissue differentiation could lead to sensitive and comprehensive prenatal diagnostic approaches to detect fetal genetic abnormalities. The same database linked to advances in gene therapy might even lead to ability to correct these abnormalities before birth, thereby substantially reducing infertility, pregnancy loss, and birth defects.

### Identifying Potential Teratogens

During pregnancy, women are exposed to a wide variety of potential teratogens—compounds that induce fetal abnormalities. As in the case of studying the fundamentals of human developmental biology, there is no practical or ethical way to identify teratogens or study their mechanisms of action in human embryos. However, embryonic stem cell screens can be used to identify and study environmental toxins and pharmaceuticals that could cause abnormalities in the differentiation of these cells. Such a screening system would provide value to drug-development programs by facilitating early identification of agents that have potential teratogenic properties. Today this screening is accomplished by exposing pregnant animals to drugs under development and examining the embryos for defects, at best only an approximation to human fetal development.

**Drug Toxicity Testing**

Because hES cells can serve as a source for all cells and tissues of the body, it will soon be possible to develop normal lines of cells that represent specific tissues and organs for testing the toxicity of new or existing drugs. The process of drug discovery is essentially characterized by development of a compound that is specific and potent with respect to the desired activity, without causing harmful side effects or irrelevant actions not necessary to achieve the therapeutic objective. Availability of normal human heart, skin, liver, or kidney cells would allow direct testing of compounds for toxicities against these cellular representatives of human organs well before human clinical testing.

One example of this potential is the case of liver cells. Not infrequently, drugs approved by The Food and Drug Administration are withdrawn from the market because of unanticipated liver toxicity that can be fatal. The frequency and severity of these events could be reduced or even eliminated if a human cellular equivalent of the human liver were available, allowing testing of compounds for liver toxicity before their introduction into clinical trials. As is the case with teratogen testing, only animal models are available to predict the effect of a new drug on human liver function.

**Regenerative Medicine**

Human embryonic stem cells should promote regenerative medicine in the near future not only because of their biologic properties, but also because they can be produced in large quantities in the laboratory under standard conditions. This is an important advantage over adult progenitor cells extracted from an individual, which are present in very low numbers, can be difficult to harvest reproducibly, and can differ in their properties among individual donors. Moreover, the likelihood of successful translation of hES cell technology into effective therapeutic approaches is supported by nearly twenty years of extensive work in mouse embryonic stem cells. These cells were first derived in 1981 (Evans and Kaufman 1981; Martin 1981) and have been used widely in mouse models of human diseases, as well as in demonstrating that differentiated cells derived from embryonic stem cells are functional when transplanted into animal models of disease.

## Cardiomyocytes for Heart Disease

Congestive heart failure, a common consequence of heart muscle or valve damage, affects nearly 5 million people in the United States, with 400,000 new cases diagnosed each year. In addition, about 1.5 million people each year suffer myocardial infarction, the primary cause of heart muscle damage, and about one-third of them die.

Heart muscle cells do not proliferate during adult life. When heart muscle is damaged by injury or obstructed blood flow, functional tissue is replaced by nonfunctional scar. Although drug therapy can be effective for some patients with congestive heart failure, inevitably the disease progresses beyond the ability of pharmacologic intervention to maintain adequate cardiac output. The fundamental pathology in damaged heart muscle is loss of functional contracting cells.

Mouse cardiomyocytes were derived from murine embryonic stem cells, purified, and injected into the hearts of recipient adult mice (Klug et al. 1996, 98). The injected cardiomyocytes repopulated myocardial tissue and stably integrated with it. This suggests that development of hES cell-derived cardiomyocytes for therapy of congestive heart failure and myocardial infarction in humans is technically feasible.

Scientists at Geron extended these observations by generating human cardiomyocytes from hES cells. These cardiomyocytes spontaneously contract in culture and express molecular markers that unequivocally define them to be human heart muscle cells (Gold et al. 2001). Much work remains before cardiomyocytes can be scaled and purified for animal and human testing, but early results suggest that, as is the case for the mouse, they can be derived and manufactured from hES cells.

## Islet Cells for Diabetes

Approximately 1.4 million Americans have insulin-dependent diabetes mellitus. They are required to take insulin injections for life to maintain normal glucose balance. However, even daily insulin injections do not prevent the secondary systemic consequences of diabetes (blindness, kidney failure, nerve damage, skin ulcers, etc). Researchers showed that mouse embryonic stem cells can produce functional insulin-secreting cells that, when purified and transferred to diabetic animals, restored normal glucose balance within a week and normal body weight within a month (Soria et al. 2000). The results strongly suggest that cell therapy with insulin-producing cells derived in similar ways from hES cells could achieve permanent cure of insulin-dependent diabetes mellitus.

## Neural Cells for Neurologic Disease

Perhaps the most near-term clinical application of hES cells lies in the treatment of neurologic disease. The opportunity to apply these cells clinically is very large: over 1 million individuals in the United States suffer from Parkinson's disease, 500,000 experience a stroke each year, and over 4 million have Alzheimer's disease. Various types of neural cells can be generated from mouse embryonic stem cells and these neural cells, when injected into sites of damage in the central nervous system (Deacon et al. 1998; McDonald et al. 1999, 5), integrate with damaged tissue and partially restore lost function. Neurons that produce dopamine (cells damaged in Parkinson's disease) have been generated from mouse embryonic stem cells (Lee et al. 2000). In animal experiments, neural cells derived from mouse embryonic stem cells and injected into sites of spinal cord damage appropriately integrated into the damaged area and resulted in partial recovery of paralysis (McDonald et al. 1999). These extraordinary findings support the application of neural cells derived from hES cells for treatment of spinal cord injuries, stroke, and potentially even Alzheimer's disease. Scientists at Geron also were successful in deriving neurons as well as neural supporting cells (astrocytes and oligodendrocytes) from hES cells (Gold et al. 2001). They are now injecting these cells into animal models of disease to try to replicate findings produced by others using mouse embryonic stem cell-derived neural cells. Much remains to be learned in terms of scale-up, purification, and control of the manufacturing process of human neural cells derived from hES cells, but early success with these cells and encouraging animal results support applications in the treatment of human neurodegenerative diseases.

## Other Medical Applications

The list of other potential applications for these cells is long and includes blood-forming stem cells to restore the hematopoietic system of patients with cancer; endothelial or blood vessel-forming cells to treat atherosclerosis, a condition that contributes to over 650,000 deaths annually in the United States; fibroblasts and keratinocyte skin cells that could be used for wound healing and treatment of burns; and chondrocytes or cartilage-forming cells that could replace cartilage for the over 16 million Americans with osteoarthritis or the over 2 million with rheumatoid arthritis.

## Problems Yet to Be Solved

These early studies are encouraging, yet many problems remain. Efforts at Geron are devoted to improving the efficiency of stem cell culturing and scale-up, optimizing conditions for manufacturing differentiated cells from hES cells, and developing methodologies to purify differentiated cells and modify them genetically to enhance their therapeutic utility. In collaboration with Celera Genomics, a PE Corporation, Geron scientists are defining gene expression profiles of undifferentiated hES cells and their differentiated daughter cells to understand what genes are important in controlling differentiation. Once these genes are identified, they can be used efficiently and naturally to cause hES cells to differentiate down desired pathways to produce therapeutically effective cells.

Other technologies will have to be developed to help translate the potential of these cells into reality. Human embryonic stem cells indefinitely renew themselves in culture in the undifferentiated state, but when they differentiate into functional neurons, liver cells, or cardiomyocytes, their replicative potential becomes finite, thereby limiting the numbers of therapeutic cells that can be produced. These undifferentiated cells are self-renewing because they produce the enzyme telomerase, which resets the normal mechanism that limits cell division, thereby allowing them to divide forever. However, this enzyme is turned off as the embryonic stem cells differentiate into functional tissue, limiting the quantity of differentiated cells that could be produced. Activation of telomerase in over ten normal human cell types confers replicative immortality to them (Bodnar 1998). Activation of the telomerase gene in the manufacturing process will allow production of limitless quantities of differentiated cells for clinical studies.

Another problem relates to immune rejection of transplanted differentiated cells. Embryonic stem cells contain marker molecules on their surfaces that are recognized by the human immune system, allowing their rejection after transplantation. Therefore, just like whole-organ transplantation of kidneys or hearts, tissue matching and immunosuppressive strategies would be required to control rejection of transplanted cells. Geron is exploring nuclear transfer technologies to produce embryonic stem cells from a patient's own tissues. Functional cells achieved in this way would escape immune rejection, thereby eliminating the need for toxic immunosuppressive drugs and tissue donors.

### Human Embryonic Stem Cells and Society

There is more to the hES cell story than chapters covering scientific themes. These cells are derived from early human embryos that, for many people, carry moral status. Like many new technologies, successful development and use of the cells for human therapeutics will depend not only on their safety and efficacy, but also on their acceptability to society at large. Although Geron's Ethics Advisory Board (Lebacqz 1999), the National Bioethics Advisory Commission (1999), and the American Association for the Advancement of Science (Chapman 1999) published suggested guidelines for ethical development of therapies based on these cells, the debate is not over.

Modern societies have the obligation to choose which alternative technologies they wish to support to improve their lives. Our hope at Geron is that after a thorough examination of the issues, many of which are explored in this volume, most people will support continued development of the technology, as do patient advocacy groups, bioethics boards, and the medical and scientific communities generally. We believe that not to develop the technology would do great harm to over 100 million patients in the United States alone who are affected by diseases potentially treatable by the many medical applications of hES cells.

## References

Amit, M., Carpenter, M. K., Inokuma, M. S., Chiu, C. P., Harris, C. P., Waknitz, M. A., Itskovitz-Eldor, J., and Thomson, J. A. 2000. Clonally derived human embryonic stem cell lines maintain pluripotency and proliferative potential for prolonged periods of culture. *Developmental Biology* 227: 271–278.

Bjornson, C. R. R., Rietze, R. L., Reynolds, B. A., Magli, M. C., and Vescovi, A. L. 1999. Turning brain into blood: A hematopoietic fate adopted by adult neural stem cells in vivo. *Science* 283: 534–537.

Bodnar, A., Ouellette, M., Frolkis, M., Holt, S. E., Chiu, C. P., Morin, G. B., Harley, C. B., Shay, J. W., Lichsteiner S., and Wright, W. E. 1998. Extension of life-span by introduction of telomerase into normal human cells. *Science* 279: 349–352.

Chapman, A. R., Frankel, M. S., and Garfinkel, M. S. 1999. Stem cell research and applications: Monitoring the frontiers of biomedical research. *AAAS/ICS Report*.

Deacon, T., Dinsmore, J., Costantini, L. C., Ratliff, J., and Isacson, O. 1998. Blastula-stage stem cells can differentiate into dopaminergic and serotonergic neurons after transplantation. *Experimental Neurology* 149: 28–41.

Evans, M. J. and Kaufman, M. H. 1981. Establishment in culture of pluripotential cells from mouse embryos. *Nature* 292: 154.

Geron Ethics Advisory Board (Lebacqz, K., Mendioloa, M. M., Peters, T., Young, E. W. D., and Zoloth-Dorfman, L.) 1999. Research with human embryonic stem cells: Ethical considerations. *Hastings Center Report* 29: 31–36.

Gold, J. D., Funk, W. D., Rosler, E. S., Inokuma, M. S., Xu, C., Thomson, J. A., Chiu, C. P., and Carpenter, M. K. 2001. Molecular and immunocytochemical analysis of human embryonic stem cell differentiation. *Developmental Biology*, manuscript under review.

Klug, M. G., Soonpaa, M. H., Koh, G. Y., and Field, L. J. 1996. Genetically selected cardiomyocytes from differentiating embryonic stem cells form stable intracardiac grafts. *Journal of Clinical Investigation* 98: 216–224.

Lee, S. H., Lumelsky, N., Studer, L., Auerbach, J. M., and McKay, R. D. 2000. Efficient generation of midbrain and hindbrain neurons from mouse embryonic stem cells, *Nature Biotechnology* 18: 675–679.

Martin, G. B. 1981. Isolation of a pluripotent cell line from early mouse embryos cultured in medium conditioned by teratocarcinoma stem cells. *Proceedings of the National Academy of Sciences USA* 78: 7634.

McDonald, J. W., Liu, X. Z., Qu, Y., Liu, S., Mickey, S. K., Turetsky, D., and Gott, D. I. 1999. Transplanted embryonic stem cells survive, differentiate and promote recovery in injured rat spinal cord. *Nature Medicine* 5: 1410–1412.

McLaren, A. 2000. Cloning: Pathways to a pluripotent future. *Science* 288: 1175–1780.

National Bioethics Advisory Commission. 1999. *Ethical Issues in Human Stem Cell Research.*

Petersen, B. E., Bowen, W. C., Patrene, K. D., Mars, W. M., Sullivan, A. K., Murase, N., Boggs, S. S., Greenberger, J. S., and Goff, J. P. 1999. Bone marrow as a potential source of hepatic oval cells. *Science* 284: 1168–1170.

Soria, B., Roche, E., Berna, G., Leon-Quinto, T., Reig, J. A., and Martin, F. 2000. Insulinsecreting cells derived from embryonic stem cells normalize glycemia in streptozotocin-induced diabetic mice. *Diabetes* 49: 157–162.

Thomson, J. 1998. Embryonic stem cell lines derived from human blastocysts. *Science* 282: 1145–1147.

# 2

## Human Embryonic Stem Cells

James A. Thomson

Human embryonic stem (hES) cells capture the imagination because they are immortal and have an almost unlimited developmental potential. After many months growing in culture dishes, these rather nondescript cells maintain the ability to form cells ranging from muscle to nerve to blood, and potentially any cell type that makes up the body. Their proliferative and developmental potential promises an essentially unlimited supply of specific cell types for transplantation in disorders ranging from heart disease to Parkinson's disease to leukemia.

### The Basic Science

To understand hES cells, it is necessary to understand something about the basic properties of early human embryos (figure 2.1). Fertilization normally occurs in the oviduct, and during the next few days a series of cleavage divisions occurs as the embryo migrates down the oviduct and into the uterus. All of the cells (blastomeres) of these cleavage-stage embryos are undifferentiated; that is, they do not look or act like the specialized cells of the adult, and the blastomeres are not committed to becoming any particular type of differentiated cell. Indeed, each blastomere has the potential to form any cell of the body. The first differentiation event occurs at about five days of development when an outer layer of cells committed to becoming part of the placenta (trophectoderm) separates from the inner cell mass (ICM). The ICM cells maintain the potential to form any cell type of the body. Because an isolated ICM lacks the trophectoderm layer, which mediates implantation, it would not be able to develop into a child if transferred to a woman's uterus.

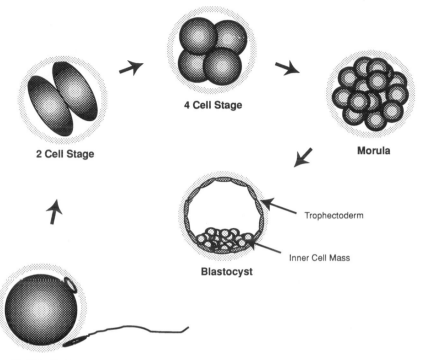

**4 Cell Stage**

**2 Cell Stage**

**Morula**

Trophectoderm

Inner Cell Mass

**Blastocyst**

**Fertilization**

**Figure 2.1**
Human preimplantation development. After fertilization, the one-cell embryo undergoes a series of cleavage divisions and forms a blastocyst at about six days of development. The blastocyst is composed of an inner cell mass and a trophectoderm. Embryonic stem cells are derived from the inner cell mass.

A hallmark of these early mammalian embryonic stages is a remarkable developmental plasticity. If a preimplantation embryo is separated into halves, each half has the ability to develop normally to term. Because of the embryo's self-regulatory abilities, the resulting twins are born a normal size and have normal life expectancies. Conversely, if two separate cleavage-stage embryos are pushed together, blastomeres can intermingle to form a single embryo that can develop to term. Such an individual would be composed of cells with different genotypes and, indeed, could have four different parents. Nonetheless, the individual could be completely normal. Thus, because of their plasticity, the concept of "individual" as applied to the adult does not apply in a straightforward way to early mammalian embryos. It is the developmental plasticity of early mammalian embryos that allows the derivation of embryonic stem (ES) cells.

A stem cell replaces itself through proliferation for prolonged periods (self-renewal) and gives rise to one or more differentiated cell types. In the adult, tissue-specific stem cells sustain tissues with a high turnover rate, such as blood, intestinal epithelium, and skin. In these tissues there is a careful balance among stem cell self-renewal, differentiation, and cell death so that tissues remain in a steady state. Adult stem cells are restricted to forming only a limited number of cell types, and some tissues, such as the heart, completely lack stem cells. In the intact embryo, cells of the ICM have the potential to form any cell type of the body, but they proliferate and replace themselves only briefly. After implantation, ICM cells differentiate to other cell types with a more restricted developmental potential. Thus, in the intact embryo, ICM cells function as precursor cells, but not as stem cells. If mammalian development were very rigid, with developmental decisions inflexibly tied to a specified number of cell divisions, ICM cells placed in culture would also just differentiate to more restricted lineages and not replace themselves, regardless of culture conditions. However, because of developmental plasticity of mammalian embryos, if the ICM is taken out of its normal embryonic environment and cultured under appropriate conditions, ICM-derived cells can proliferate and replace themselves indefinitely, yet maintain the developmental potential to form any cell type. These pluripotent, ICM-derived cells are ES cells.

The derivation of ES cell lines from mouse blastocysts was first reported by two independent groups in 1981 (Evans and Kaufman 1981; Martin 1981). The term "ES" cell was introduced to distinguish the origin of these cells from the origin of embryonal carcinoma (EC) cells, which are pluripotent stem cell lines derived from teratocarcinomas. Teratocarcinomas are malignant germ cell tumors that include a mixture of different kinds of differentiated cells. Mouse EC cell lines were used as an in vitro model of mammalian differentiation for years before the derivation of mouse ES cell lines, and the derivation of pluripotent human EC cells was reported in the early 1980s (Andrews et al. 1984; Damjanov and Solter 1976). However, possibly because of their origin in the malignant tumor environment, EC cell lines generally have a much more restricted developmental potential than ES cell lines. Human EC cell lines, for example, have severe chromosomal abnormalities and a fairly limited developmental potential (Roach et al. 1993). Mouse ES cells injected into an intact preimplantation embryo can intermingle with the host embryo and contribute to normal development, forming a chimera. The ES cells can contribute to any tissue of the chimera, including germ cells, which gives developmental biologists a method to manipulate the germ line of mice. Pluripotent cell lines similar to mouse ES cells have been derived from primordial germ cells, cells that would normally develop into either sperm or egg (Matsui et al. 1992; Resnick et al. 1992). Again, to distinguish their origin, these pluripotent cell lines are referred to as embryonic germ (EG) cell lines. Human EG cell lines were recently derived from human fetal germ cells (Shamblott et al. 1998).

Human ES cell lines were derived from blastocyst-stage preimplantation embryos produced by in vitro fertilization (Thomson et al. 1998). To derive hES cell lines, the ICM of the blastocyst is isolated from the trophectoderm layer and plated on mouse embryonic fibroblasts. After approximately two weeks of culture, ICM-derived cells are dissociated and replated. Undifferentiated hES cells have a characteristic morphology that includes a high nuclear:cytoplasmic ratio and numerous prominent nucleoli (figure 2.2). Human ES cells, human EC, and nonhuman primate ICM cells all express characteristic cell surface markers, including stage-specific embryonic antigen, that differ from those expressed by mouse ES cells (Andrews et al. 1984a,b, 1987; Kannagi et al. 1983; Solter and Knowles 1978; Wenk et al. 1994). The shared pattern of expression of cell surface markers by ES cells, human EC cells, and nonhuman primate ES cells, and different pattern of expression by mouse ES cells, reflects fundamental embryologic differences between primates and mice.

The hES cell lines derived to date have a normal complement of chromosomes and are capable of prolonged proliferation. Normal (diploid) human somatic cells proliferate in culture for a characteristic number of times and then stop dividing (replicative senescence). Neoplastic somatic human cells that escape this "mortality" invariably have significant chromosomal changes. Because hES cell lines are derived from very early

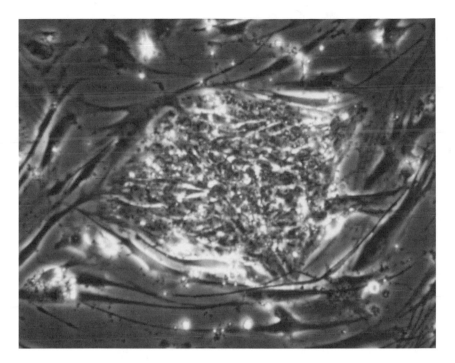

**Figure 2.2**
Human embryonic stem cells on mouse fibroblasts.

embryos, they naturally express high levels of the enzyme associated with cellular immortality, telomerase. No hES cell line has been observed to undergo replicative senescence, and one line was cultured continuously for well over a year and maintained a normal karyotype, suggesting that these cell lines are capable of unlimited proliferation. Because proliferation of undifferentiated hES cells appears to be unlimited, it should be possible that unlimited numbers of differentiated derivatives could one day be produced in culture.

When removed from fibroblast feeder layers, hES cells differentiate into a variety of cell types. Leukemia-inhibitory factor, which prevents differentiation of mouse ES cells, fails to prevent differentiation of hES cells in the absence of fibroblasts. Because production of fibroblast feeder layers is labor intensive, the amount of ES cells that can be grown would not yet be therapeutically useful. Thus, identification and purification of factors produced by fibroblast feeder layers that sustain the undifferentiated proliferation of hES cells is a critical research area, because replacing fibroblasts with purified factors would allow routine large-scale production of hES cells.

When hES cells are allowed to differentiate in the absence of fibroblasts, they differentiate into a variety of cell types including endoderm, neural cells, and muscle cells. When they are injected into immunocompromised mice, they form teratomas with differentiation of several cell types, including ciliated respiratory epithelium and gut epithelium (endoderm); striated muscle, smooth muscle, cartilage, bone, and connective tissue (mesoderm); neural tissue, skin, and hair (ectoderm); and numerous unidentified types (figure 2.3). Within hES cell teratomas there is abundant evidence of coordinated interactions among cells, and even among cells originating from different embryonic germ layers. For example, development of hair requires coordinated interactions between the overlying ectoderm and underlying mesenchyme.

Because of possible harm to the resulting child, it is not ethically acceptable to manipulate the postimplantation human embryo experimentally, so we are largely ignorant about the mechanisms of early human embryology. Most of what is known about human development, especially in the early postimplantation period, is based on histologic sections of limited numbers of human embryos and on analogy to mouse embryology. However, human and mouse embryos differ significantly, particularly in the formation, structure, and function of fetal membranes and placenta, and formation of an embryonic disk instead of an egg cylinder (Benirschke and Kaufmann 1990; Luckett 1975, 1978). For example, the mouse yolk sac is a well-vascularized, robust, extraembryonic organ throughout gestation and has important nutrient exchange functions. In humans, the yolk sac has important early functions, including initiation of hematopoiesis and germ cell migration, but later in gestation it is essentially a vestigial structure. Similarly, dramatic differences exist between mouse and human placentas, both in structure and function. Thus, for understanding developmental events that support the initiation and maintenance of human pregnancy, mice can provide only limited understanding. The hES cell lines provide an important new in vitro model that will improve our understanding of the differentiation of

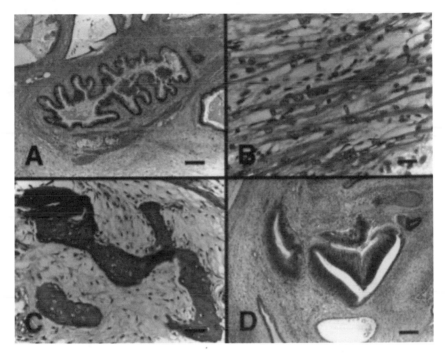

**Figure 2.3**
Differentiation of human embryonic stem cells in teratomas. (A) gut; (B) striated muscle; (C) bone; (D) neural epithelium.

human tissues and thus provide important insights into such processes as infertility, pregnancy loss, and birth defects.

## Implications and Importance of the Research

As developmental biologists become more accomplished at directing hES cells to specific cell types, the differentiated derivatives of the cells should have an important role in developing new therapies. Large, purified populations of hES cell-derived cells, such as heart muscle cells or neurons, could be used to screen for new drugs. Purified, normal human cells would allow accurate screening of candidate drugs, greatly reduce the need for animal testing during the early screening process, and accelerate drug discovery. Differentiated derivatives of hES cells also could be used to test for possible toxic side effects of drugs identified by other methods, and hES cells would be particularly useful for identifying compounds that interfere with normal development.

Finally, differentiated derivatives of hES cells could be applied to transplantation therapies for treatment of a range of human diseases. Because certain diseases result from the death or dysfunction of just one or a few cell types, replacing those cells by transplantation could offer long-term treatment. The hES cells can proliferate indefinitely and differentiate to many, perhaps all, cells of the body. Therefore they have the potential to provide a limitless source of specific cell types for transplantation. Numerous diseases might be treated by this approach, including heart disease, juvenile-onset diabetes, Parkinson's disease, and leukemia. Developmental biology has made dramatic strides in recent years, but it is not yet possible to direct ES cells efficiently to most specific cell types. Significant progress has been made, however, in differentiating the cells to specific lineages, including blood, neural, and muscle cells (Brustle et al. 1999, 1997; Keller 1995; Klug et al. 1996).

Introducing ES cell-derived cells into the body so that they restore useful function to a damaged organ and preventing their rejection are problems likely to prove even more challenging than deriving specific cell types. Cell types whose function require coordinated, three-dimensional integration into host tissue will prove particularly challenging. For example, in a heart attack, part of the heart muscle dies because of a blockage of the blood supply to the muscle. Because an adult has no heart muscle stem cell, if the patient survives, dead heart muscle is permanently replaced by nonfunctional scar tissue. Human ES cells spontaneously differentiate to heart muscle cells in tissue culture, and it should be possible to purify those cells from other cell types. However, getting them back into a damaged heart to replace dead muscle or scar tissue and actually restoring function to the heart will prove challenging. Not only must new muscle integrate in a mechanically useful way with surrounding muscle, but new blood vessels will be required to supply the new muscle or it will die. Progress in the field of angiogenesis suggests that inducing new growth of existing vessels to supply transplanted heart muscle may one day be possible, but considerable research is required before this is practical.

After transplanted tissue is successfully integrated, its rejection by the patient's immune system must be prevented. Possible strategies include banks of major histocompatibility complex-typed hES cell lines, genetically modified hES cell lines that are designed to be less immunogenetic, and ES cells genetically identical to a specific patient produced by nuclear transfer. Nuclear transfer technology offers potentially the most effective and most controversial solution. For human medicine, the profound implication of the cloning of Dolly (Wilmut et al. 1997) is that development may be more flexible than once thought, and differentiated cells can be reprogrammed into undifferentiated cells. Dolly was cloned by transplanting the nucleus from a mammary epithelial cell to an enucleated oocyte, and by transferring the resulting nuclear transfer product to a recipient ewe. The same procedure could be performed with a human somatic cell nucleus transferred to an enucleated human oocyte, but instead of transferring the nuclear transfer product to produce a pregnancy, a blastocyst could be produced in vitro and an ES cell line derived. Through this method it might be possible to take a readily accessible cell type such as a skin fibroblast from a biopsy specimen, establish an ES cell line using nuclear transfer from the fibroblast, direct the cell line to heart muscle cells, and transplant those heart muscle cells back to the patient who donated the fibroblast. The heart muscle cells would be genetically identical to the patient's cells for all nuclear-encoded genes. Production of a human embryo by nuclear transfer for therapeutic purposes would be extremely controversial. Reprogramming a differentiated cell nucleus by human oocyte cytoplasm to create an ES cell line has not been demonstrated so it is not even certain that human oocyte cytoplasm has this ability.

Because hES cell-based transplantation therapies are new and unproved, it will be essential to demonstrate their safety and efficacy in an accurate animal model. Rhesus monkey ES cells are very similar to hES cells and rhesus monkeys share a close evolutionary and physiologic relationship with humans (Thomson et al. 1995; Thomson and Marshall 1998). Several important diseases that might be treated by ES cell-based therapies, including Parkinson's disease and diabetes mellitus, have accurate rhesus monkey models (Burns et al. 1983; Jones et al. 1980). Nuclear transfer techniques have been developed in the rhesus monkey (Meng et al. 1997). Elucidating the basic molecular mechanisms by which the oocyte reprograms adult nuclei in this primate species may one day allow the direct reprogramming of human nuclei to produce an ES cell line without having to produce an embryo as an intermediate step. If it becomes possible to derive an ES cell line from a source other than an embryo, ethical controversies surrounding hES cells would greatly diminish.

## References

Andrews, P. W., Banting, G., Damjanov, I., Arnaud, D., and Avner, P. 1984a. Three monoclonal antibodies defining distinct differentiation antigens associated with different high molecular weight polypeptides on the surface of human embryonal carcinoma cells. *Hybridoma* 3: 347–361.

Andrews, P. W., Damjanov, I., Simon, D., Banting, G., Carlin, C., Dracopoli, N., and Fogh, J. 1984b. Pluripotent embryonal carcinoma clones derived from the human teratocarcinoma cell line Tera-2. *Laboratory Investigation* 50: 147–162.

Andrews, P. W., Oosterhuis, J., and Damjanov, I. 1987. Cell lines from human germ cell tumors. In: Robertson, E., ed. *Teratocarcinomas and Embryonic Stem Cells: A Practical Approach.* Oxford: 26 IRL Press, pp. 207–246.

Benirschke, K. and Kaufmann, P. 1990. *Pathology of the Human Placenta.* New York: Springer-Verlag.

Brüstle, O., Spiro, A. C., Karran, K., Choudhary, K., O'Kabe, S., and McKay, R. D. G. 1997. In vitro generated neural precursors participate in mammalian brain development. *Proceedings of the National Academy of Sciences of the USA* 94: 14809–14814.

Brüstle, O., Jones, K. N., Learish, R. D., Karram, K., Choudhary, K., Wiestler, O. D., Duncan, I. D., and McKay, R. D. G. 1999. Embryonic stem cell-derived glial precursors: A source of myelinating transplants. *Science* 285(75): 1–753.

Burns, R. S., Chiueh, C. C., Markey, S. P., Ebert, M. H., Jacobowitz, D. M., and Kopin, I. J. 1983. A primate model of parkinsonism: Selective destruction of dopaminergic neurons in the pars compacta of the substantia nigra by N-methyl-4-phenyl-1,2,3,6-tetrahydropyridine. *Proceedings of the National Academy of Sciences of the USA* 80: 4546–4550.

Damjanov, I. and Solter, D. 1976. Animal model of human disease: Teratoma and teratocarcinoma. *American Journal of Pathology* 83: 241–244.

Evans, M. and Kaufman, M. 1981. Establishment in culture of pluripotential cells from mouse embryos. *Nature* 292: 154–156.

Jones, C. W., Reynolds, W. A., and Hoganson, G. E. 1980. Streptozotocin diabetes in the monkey: Plasma levels of glucose, insulin, glucagon, and somatostatin, with corresponding morphometric analysis of islet endocrine cells. *Diabetes* 29: 536–546.

Kannagi, R., Cochran, N. A., Ishigami, F., Hakomori, S., Andrews, P. W., Knowles, B. B., and Solter, D. 1983. Stage-specific embryonic antigens (SSEA-3 and -4) are epitopes of a unique globo-series ganglioside isolated from human teratocarcinoma cells. *EMBO Journal* 2: 2355–2361.

Keller, G. M. 1995. In vitro differentiation of embryonic stem cells. *Current Opinion in Cell Biology* 7: 862–869.

Klug, M. G., Soonpaa, M. H., Koh, G. Y., and Field, L. J. 1996. Genetically selected cardiomyocytes from differentiating embronic stem cells form stable intracardiac grafts. *Journal of Clinical Investigation* 98: 216–224.

Luckett, W. P. 1975. The development of primordial and definitive amniotic cavities in early rhesus monkey and human embryos. *American Journal of Anatomy* 144: 149–168.

Luckett, W. P. 1978. Origin and differentiation of the yolk sac and extraembryonic mesoderm in presomite human and rhesus monkey embryos. *American Journal of Anatomy* 152: 59–98.

Martin, G. 1981. Isolation of a pluripotent cell line from early mouse embryos cultured in medium conditioned by teratocarcinoma stem cells. *Proceedings of the National Academy of Sciences of the USA* 78: 7634–7638.

Matsui, Y., Zsebo, K., and Hogan, B. L. 1992. Derivation of pluripotential embryonic stem cells from murine primordial germ cells in culture. *Cell* 70: 841–847.

Meng, L., Ely, J. J., Stouffer, R. L., and Wolf, D. P. 1997. Rhesus monkeys produced by nuclear transfer. *Biology of Reproduction* 57: 454–459.

Resnick, J. L., Bixler, L. S., Cheng, L., and Donovan, P. J. 1992. Long-term proliferation of mouse primordial germ cells in culture. *Nature* 359: 550–551.

Roach, S., Cooper, S., Bennett, W., and Pera, M. F. 1993. Cultured cell lines from human teratomas: Windows into tumour growth and differentiation and early human development. *European Urology* 23: 82–88.

Shamblott, M. J., Axelman, J., Wang, S., Bugg, E. M., Littlefield, J. W., Donovan, P. J., Blumenthal, P. D., Huggins, G. R., and Gearhart, J. D. 1998. Derivation of pluripotent stem cells from cultured human primordial germ cells. *Proceedings of the National Academy of Sciences of the USA* 95: 13726–13731.

Solter, D. and Knowles, B. B. 1978. Monoclonal antibody defining a stage-specific mouse embryonic antigen (SSEA-l). *Proceedings of the National Academy of Sciences of the USA* 75: 5565–5569.

Thomson, J. A., Liskovitz-Eldor, J., Shapiro, S. S., Waknitz, M. A., Swiergiel, J. J., Marshall, V. S., and Jones, J. J. 1998. Embryonic stem cell lines derived from human blastocysts. *Science* 282: 81145–81147.

Thomson, J. A., Kalishman, J., Gobs, T. G., Durning, M., Harris, C. P., Becker, R. A., and Heam, J. P. II. 1995. Isolation of a primate embryonic stem cell line. *Proceedings of the National Academy of Sciences of the USA* 92: 7844–7848.

Thomson, J. A. and Marshall, V. S. 1998. Primate embryonic stem cells. *Current Topics in Developmental Biology* 38: 133–165.

Wenk, J., Andrews, P. W., Casper, I., Hata, J., Pera, M. F., von Keitz, A., Damjanov, I., and Fenderson, B. A. 1994. Glycolipids of germ cell tumors: Extended gbobo-series glycolipids are a hallmark of human embryonal carcinoma cells. *International Journal of Cancer* 58: 108–115.

Wilmut, I., Schnieke, A. E., McWhir, J., Kind, A. J., and Campbell, K. H. S. 1997. Viable offspring derived from fetal and adult mammalian cells. *Nature* 385: 810–813.

# 3

# The Stem Cell Debate in Historical Context

John C. Fletcher

On November 14, 1998, President Clinton asked the National Bioethics Advisory Commission (NBAC) to provide a thorough review of all issues surrounding human stem cell research, "balancing all ethical and medical considerations." Ten months later, NBAC (1999) submitted its report built on foundations laid by other commissions and advisory panels. One goal of this chapter is to single out the contribution of the first public bioethics body that dealt with controversial issues surrounding use of fetuses and embryos in research. In 1975 the National Commission for the Protection of Human Subjects of Biomedical and Behavioral Research (National Commission) set an enduring example for how public bioethics can contribute to compassionate compromises on controversial issues. Failure to adhere to that example damages the effectiveness of NBAC's work.

In 1973 the Supreme Court ruled that a fetus is not a person in the context of constitutionally protected rights (*Roe v. Wade* 410 U.S. 113). The door was opened to freedom of choice in abortion. Members of Congress began to worry about possible exploitation of aborted fetuses.[1] The National Institutes of Health (NIH) imposed a moratorium on fetal research. In 1974 Congress established the National Commission and charged it with formulating ethical and public policy guidelines for fetal research.

The commission's report (1975), issued some four months later, was a compromise between liberal and conservative views on fetal research. In accord with liberal views, the commission encouraged fetal research because of its many benefits, such as development of polio and rubella vaccines. Yet it also sharply restricted fetal research: where research risks were concerned, fetuses to be aborted had to be treated equally with fetuses to be delivered. This bold specification of a principle of equality of protection honored a conservative viewpoint that contrasted markedly with the utilitarian ethos previously dominating United States research practices.

In coming to this conclusion, the commission drew on the work of several ethicists from conservative traditions. A leading Catholic moral theologian, Richard McCormick, stated that "the fetus is a fellow human being, and ought to be treated ... exactly as one treats a child" (1976). McCormick would permit fetal research provided there was "no discernible risk, no notable pain, no notable inconvenience, and ... promise of considerable benefit" (8). His term "no discernible risk" later evolved into the category of "minimal risk." The meaning of this term continues to be controversial and widely challenged. McCormick's position opened a way conceptually for those giving primary rights to the fetus to accept fetal research nonetheless. However, with his position came his restrictive risk standard and, most important, the underlying premise that fetuses ought to be treated equally, as "fellow human beings."

The principle of equal treatment was also picked up by LeRoy Walters, a Protestant ethicist. Walters advised the commission to use a principle of equality of protection whether fetuses were destined for abortion or for delivery. Under this Golden Rule idea, researchers could not impose a higher risk with a fetus to be aborted than they would with a fetus to be delivered (Walters 1976). Although other ethicists also influenced the commission, it was the strength of these two positions that led the commission to offer the possibility of research on fetuses to be aborted, provided the risks were minimal and were only what would be accepted for a fetus going to term.

Thus, in effect, in spite of *Roe v. Wade*, the National Commission declared that societal protection of human subjects ought to be extended to fetuses, even to those slated for abortion. Hence, any in utero fetal research, especially that not designed to benefit the health of the fetus, had to conform to a standard of minimal risk. To make this compromise work, the commission envisioned a continuing ethics advisory board

(EAB) as a resource for local institutional review boards (IRBs) and for developing national policy. Some commissioners worried that important fetal research could not be done ethically without selectively assigning higher risks to fetuses to be aborted than to those going to term. The commission invested great hope in a future EAB to make decisions on a case-by-case basis. Its report can be seen as a compromise premised on strong hopes for the work of an EAB that would function like a national IRB.

Regulations for fetal research were promulgated (45 CFR 46) and the moratorium was lifted on July 29, 1975. The regulations distinguish research to meet the health needs of the fetus from research to develop "important biomedical knowledge which cannot be obtained by other means." Only minimal risk is permitted in the latter category. Minimal risks were defined as "not greater in and of themselves than those ordinarily encountered in daily life or during the performance of routine physical or psychological examinations or tests" (45 CFR 46.102i). Application of such a standard to fetuses has never been clarified. Furthermore, the commission envisioned the possibility of occasional waivers approved by an EAB, and indeed one such waiver was granted (Steinfels 1979). However, the charter of the EAB lapsed in 1980 and was not renewed; there has been no EAB since.

After the 1984 election when President Reagan was retained in office, Congress enacted legislation far more protectionist than federal regulations that followed the work of the National Commission. Public law 99–158 effectively nullified the minimal risk standard and ended federal support of fetal research involving any level of risk, including into normal fetal physiology (Fletcher and Schulman 1985). The cost to the nation's health, especially to the health of children, is difficult to calculate but potentially enormous.

Two other laws had a significant impact on the situation in which stem cell research emerged. In 1993 Congress lifted a moratorium on federal funding of in vitro fertilization (IVF) research and nullified the requirement for EAB approval of such research. In 1996 a new Congress banned federal funding for embryo research and dashed NIH hopes to fund improvements of IVF and other projects involving human embryos. The Human Embryo Research Panel had argued in 1994 for federal funding for this research. However, even before the official ban, the threat of strong opposition from Congress toward any embryo research inhibited NIH approval of several clinically relevant projects that had passed NIH scientific review (Charo 1995). After the ban, the NIH received no proposals involving embryo research. Foregoing NIH involvement in arenas such as cancer, genetic research, infertility, and contraceptive research entails large losses to science and costs to human health. These losses ought to arouse moral concern, as obligations of beneficence and utility cannot be met without improvements in maternal and fetal health.

The premise of the federal ban is that embryos, like fetuses, deserve virtually absolute societal protection from destruction or harm in research activities. The language of the ban ("risk of injury or death") is based on earlier federal law restricting funding for fetal research. Congress also prohibited federal funding (with three exceptions) for elective abortions in the Medicaid program.[2] Conservative views of the moral status of the embryo and fetus prevail in the federal sector of science but stop at the border of the private sector.

Private sector research is constrained only by state laws prohibiting embryo research. In states with no laws against it, the research is essentially unregulated. The legality of stem cell research in various states is therefore a complex topic. As Andrews (1999) noted, twenty-four states have no laws specifically addressing research on embryos and fetuses, but in these states legal precedents regarding privacy, informed consent, and commercialization may come into play.

Researchers thus work in a morally bifurcated universe: prohibitive in the public sector and permissive in the private sector. The abortion issue is so explosive that social equilibrium requires such a morally divided universe. In a democracy, the elected majority's beliefs can prevail when Congress denies funding for activities it deems immoral. Federal and state governments may also use denial of funding to cool the heat of moral disputes around topics such as abortion.

In this political climate, it is no surprise that NBAC takes conservative moral opinion very seriously. Its chapter on ethical issues begins with a lengthy and thoughtful response to moral objections to using fetal tissue to derive stem cells for human embryonic germ (hEG) cell research. The argument builds on the work of the NIH Human Fetal Tissue Transplantation Panel (1988) but does not assume that the moral case was decided by that panel. Two objections in particular required new response: that providers of fetal tissue for hEG research are morally "complicit" with the preceding abortion, and that researchers are causally responsible for abortions that women can choose with an easier conscience because they believe that others may benefit.

To these objections, NBAC gave three responses. First, no data show that fetal tissue research increases the abortion rate; this weakens the claim that the research contributes to abortion. Second, legal safeguards in effect since 1993 protect against abuses by ensuring that women's consent for abortion must precede any request for consent to fetal tissue research, that women receive no payment for fetal tissue, that no alterations be permitted in the timing or procedures used in performing the abortion, and so on. Third, NBAC maintains that if providers of fetal tissue are to be held causally responsible for abortions, many others would also have to be held causally responsible; for example, those who encourage women to seek education would be responsible if a woman had an abortion in order to continue her education. Since no one holds people causally responsible in such instances, NBAC found that researchers also should not be deemed morally responsible for women's choice to abort.

The case of human embryonic stem (hES) cell research using excess embryos is similar in every respect except for the fact that researchers' actions cause embryos to die. Here, NBAC framed its moral position largely in terms of loyalty to medicine's goals of healing, prevention, and research. These goals were envisioned as "rightly characterized by the principles of beneficence and nonmaleficence"—doing good and avoiding harm. Thus, NBAC drew on widely accepted principles and also on a balance of goods and harms that echoes utilitarian reasoning. A benefit:harms ratio is a familiar tool in ethics to weigh and balance foreseeable consequences of actions. I will return to an analysis of NBAC's position itself in another chapter. What is crucial here is that, by drawing on several modes of reasoning, NBAC may be trying to provide a compromise, following in the footsteps of the National Commission so many years ago. Whether it is as successful in that compromise as was the commission is another question.

At some points, NBAC appears to have worked hard for compromise between liberals and conservatives. It describes clashing views on the question of the moral status of the embryo, and intends to be respectful of all reasonable alternative views. Here is a genuine search for common ground between liberals and conservatives. Noting that conservatives often make room for *some* instances of abortion, and that this suggests *some* grounds on which fetal life can be taken, NBAC suggested that such conservatives might willingly cross the gap to permit hES research to save lives or prevent disability, especially if adequate safeguards are in place.

However, in the final analysis, NBAC gave up the search for compromise. When the analogy between permissible abortion and research on hES cells broke down, NBAC turned to urging a benefit:harm ratio. Ultimately, it took the position that embryos are forms of human life but not human subjects of research. Whereas the couple who donate gametes are clearly subjects for purposes of research conducted on their embryos, the embryos themselves are not yet fully subjects. In short, NBAC took a stand on the moral status of the embryo, but simply asserted this stand and did not provide convincing argument for it. The stand is similar to that taken by the EAB in 1979 and by the Human Embryo Research Panel in 1994: an embryo merits respect as a form of human life, but not the same level of respect as would be accorded to persons. This is not a compromise with those who hold that the embryo is a person.

All hope for moral compromise disappeared when NBAC moved to recommend permitting federal funding to derive hES cells from excess embryos. Eight reasons were given to support this position. They ranged from the importance of science to the need for federal support and regulation to avoid industry-driven research, which of necessity operates with some secrecy and limits dissemination of results. These reasons are plausible in themselves; however, the issue is the politics of embryo research. Whose political interests does the NBAC's stand serve? Here, the position on federal funding requires conservatives to compromise their moral beliefs while liberals compromise nothing. This is not a true compromise.

Let us assume for purpose of argument that NBAC is morally right in holding that the ban should be amended to permit federal support for stem cell research. But then let us assume that it is not politically possible to amend the ban now. Congress does not now have a majority who would enact the NBAC position. Hence, what ought to be done cannot be done. In this context, the stage is set for genuine compromise between liberals and conservatives that facilitates the appropriations process for federal funding for embryo research using excess IVF embryos. What might that compromise be?

Given the history of prohibitive policies of Congress and a great need for public education on these issues, one can assume first that Congress and the public will be more easily persuaded to amend the ban if the focus of federal funding is on therapeutic aims rather than on basic research, and second, that arguments such as those made by NBAC regarding the great potential good of this research will become more persuasive as basic research matures and we enter a stage of readiness for clinical trials. A reasonable compromise would be to defer amending the ban until that point is reached.

This will seem unfair to the liberal mind. However, if one takes the moral opinion of conservatives seriously, as did the NBAC in part of its ethical stance, it follows that federal funding for the derivation of ES cells from excess embryos ought to be a last resort to mount research aimed at saving lives and preventing disability. This may not move the research agenda forward as quickly as liberals would like, but it is a fair compromise with the conservative position that would ban forever all federal support for this research.

The NBAC was charged with balancing medical and ethical considerations. Its work is in fact unbalanced because it did not follow the example of the National Commission and allow the moral logic of compassionate compromise to guide its choices and recommendations on federal funding. Public bioethics is unavoidably political because it aims to influence public policy. In this context, more attention should have been given to the history of congressional actions on fetal and embryo research, the political weight of conservative moral views, and a volatile political context. Public bioethics must be concerned with the politically possible in order to achieve the right balance between competing factors, especially on such a controversial issue. The NBAC's ethical analysis appeals to liberal thought but disappoints pragmatists in ethics and politics. Most unfortunate, it seriously distances this decade's public bioethics body from conservative moral opinion.

## Notes

1. A very informative political history of events before 1988 is found in Lehrman (1988).

2. First introduced in 1976, the Hyde Amendment, named for its sponsor, Henry Hyde (R.-Il.), restricts all funding of abortion for the federal share of Medicaid except for cases in which two physicians attest that continuation of the pregnancy will result in severe and lasting damage to the woman's physical health, and in cases of reported rape and incest. The law took effect after a Supreme Court ruling: *Harris v. McRae* 448 U.S. 297 (1980).

## References

Andrews, L. B. 1999. State regulation of embryo stem cell research. In National Bioethics Advisory Commission. *Ethical Issues in Human Stem Cell Research*. Vol. II. *Commissioned Papers*. Rockville, MD: National Bioethics Advisory Commission.

Charo, R. A. 1995. The hunting of the snark: The moral status of embryos, right-to-lifers, and third world women. *Stanford Law and Policy Review* 6: 11–27.

Fletcher, J. C. and Schulman, J. D. 1985. Fetal research: The state of the question. *Hastings Center Report* 15: 6–12.

Human Embryo Research Panel. 1994. *Report of the Human Embryo Research Panel*. Washington, DC: National Institutes of Health.

Lehrman, D. 1988. *Summary: Fetal Research and Fetal Tissue Research*. Washington, DC: American Association of Medical Colleges.

McCormick, R. 1976. Experimentation on the fetus: Policy proposals. In *Appendix to Report and Recommendations: Research on the Fetus*. Washington, DC: National Commission for the Protection of Human Subjects of Biomedical and Behavioral Research.

National Bioethics Advisory Commission. 1999. *Ethical Issues in Human Stem Cell Research*. Rockville, MD: National Bioethics Advisory Commission.

National Commission for the Protection of Human Subjects of Biomedical and Behavioral Research. 1975. *Report and Recommendations: Research on the Fetus*. Washington, DC: U.S. Department of Health, Education, and Welfare.

National Institutes of Health. 1994. Report of the Human Fetal Tissue Transplantation Research Panel. Vol. 1. Bethesda, MD.

Steinfels, M. 1979. At the EAB. Same members, new ethical problems. *Hastings Center Report* 5: 2.

Walters, L. 1976. Ethical and public policy issues in fetal research. In *Appendix to Report and Recommendations: Research on the Fetus*. Washington, DC: U.S. Department of Health, Education, and Welfare.

# II
## Raising the Ethical Issues

# 4

# On the Ethics and Politics of Embryonic Stem Cell Research

Erik Parens

In November 1998 President Clinton requested that the National Bioethics Advisory Commission (NBAC) turn its attention to emerging issues in embryonic stem (ES) cell research. In his response to the 1994 Human Embryo Research Panel's (HERP) report, the president had said that although he could endorse research on embryos originally created by means of in vitro fertilization (IVF) for the purpose of reproduction, he could not endorse using IVF to create embryos for research. In early 1998, however, he indicated that he could endorse using somatic cell nuclear transfer (SCNT) to create embryos for research. Indeed, the administration announced that it could not support a ban on SCNT unless the ban allowed an exception to "permit [SCNT] using human cells for the purpose of developing stem-cell technology to prevent and treat serious and life-threatening diseases" (Statement of Administration Policy, 1998).

I believe that an intellectually honest and adequate response to the president's request will acknowledge, if not fully address, the following five questions. First, how should policy makers view and talk about the relationship between human (h)ES cell research and embryo research? Is it reasonable to attempt to cordon off public conversation about the former from public conversation about the latter? Second, what is the current state of the policy conversation concerning embryo research? Specifically, what was the HERP argument for limited embryo research and how could it have been made more persuasively? Third, if in general it were acceptable to do limited research on embryos, would the original *intention* of the maker of the embryo make a moral difference? Is it reasonable to endorse research on discarded embryos, but oppose research on created ones? Fourth, if agreement were reached that under carefully circumscribed conditions it is acceptable to create embryos for research, would it make a moral difference which *means* are used to create them? Was it reasonable for the president to endorse SCNT to create embryos for research but not to endorse IVF for the same purpose? Finally, if it were acceptable to use SCNT with human cells to produce embryos for research, would it be acceptable to use SCNT with human and nonhuman cells for the same purpose?

## The Relationship between hES Cell Research and Embryo Research

The director of NIH, Harold Varmus, requested a legal opinion on whether it was permissible to use federal funds for research on hES cells. In January 1999 Dr. Varmus's counsel, Harriet Rabb, rendered an opinion, which acknowledges that federal funding may not be used for "research in which a human embryo or embryos are destroyed." Thus she acknowledges that insofar as isolating hES cells requires destroying embryos, using federal funds to isolate hES cells is prohibited. According to Ms. Rabb's reading of the law, however, insofar as hES cells themselves are not embryos, federal funding for research on them is not prohibited.[1]

This legal distinction between embryos and stem cells obscures the fact that hES cell research and embryo research are inextricably entwined. For example, when HERP discussed embryo research that was acceptable for federal funding, one area it identified was "research involving the development of [hES] cells" (1994, 10). Not only are ES cells isolated by dismantling embryos, but in principle it seems that hES cells could be transformed into embryos: in mice, they were fused with tetraploid host blastocysts to form embryos (and mature animals) that were "solely derived from the ES cells" (Solter and Gearhart 1999). In the era of somatic cell nuclear transfer, however, when potentially all somatic cells can be transformed into embryos, it may seem that the capacity of ES cells to be so transformed does not make their relationship to embryos especially significant.

That view overlooks an important characteristic of ES cells. Whereas the public policy conversation has focused on their pluripotentiality, it has largely ignored their so-called immortality, or, more accurately, their capacity for "prolonged undifferentiated proliferation" (Thomson et al. 1998). Because ES cells "grow tirelessly in culture, ... they give researchers ample time to add or delete DNA precisely" (Regalado 1998, 40). Because it is easier to make precise gene insertions in ES cells than it is to make such insertions in other kinds of cells, including zygotes and somatic cells (Gordon 1999), ES cells are potentially a powerful tool with which to produce germ line interventions.

Thus ES cells and embryos are importantly related: with relative ease, ES cells could be genetically altered, those altered cells could be fused with a disabled blastocyst to give rise to an embryo derived solely from the ES cells, and that embryo could give rise to a genetically altered organism.[2] To be sure large practical (not to mention ethical) obstacles stand in the way of using ES cells to produce germ line alterations in humans (Solter and Gearhart 1999; Gordon 1999), but it is at least theoretically possible that in the future, practical obstacles that now exist will be overcome. A comprehensive analysis of ES cell research should acknowledge this theoretical possibility. Careful analysis will avoid too quickly asserting that there is nothing special about the capacity of ES cells to be transformed into embryos.

Ms. Rabb may be accurate in saying that as ES cells are not embryos, the letter of the law against embryo research does not apply to them. However, insofar as ES cells are harvested by destroying embryos and can in principle be used to produce not just embryos but altered embryos (i.e., with added or deleted genes), and insofar as the spirit of the law aims to prevent such destruction and production, the spirit of the law does apply.

Although I believe that the current congressional ban against all embryo research is not in the public interest, I also believe that public policy makers are obliged to respect that ban or make the arguments to lift it. A legalistic end run around the spirit of the law is contrary to what we might call a basic rule of public policy: makers of such policy are obliged to speak openly and clearly about what publicly funded research entails. Medical progress is a very great good. But in a democracy, transparent and respectful public conversation may be an even greater good.

## The State of the Policy Argument about Embryo Research

The major argument for doing embryo research is that it promises to reduce human suffering and promote well-being. The major argument against using embryos for research is that they have the moral status of persons and thus should not be destroyed, no matter how great the human benefit.

In 1994 HERP rejected the position that embryos have the same moral status as persons: "That is because of the absence of developmental individuation in the preimplantation embryo, the lack of even the possibility of sentience and most other qualities considered relevant to the moral status of persons, and the very high rate of natural mortality at this stage." However, the panel did state that "the human embryo warrants serious moral consideration as a developing form of human life."

Thus HERP proposed a way between two radical alternatives. The panel could not accede to the view that embryos are persons. That view is persuasive only if one proceeds from a particular set of beliefs that citizens in a democracy are not obliged to accept. Nor, however, could the panel accede to the view that embryos are mere property. That view is persuasive only if one chooses to ignore that these entities could become human beings. In light of the determination that embryos have an intermediate moral status, being neither persons nor property (Steinbock 1994), HERP suggested that appropriate respect could be shown for embryos by limiting the time frame in which research is done on them and by limiting the purposes to which they can be put. This is as reasonable a recommendation as any policy group is likely to make. Nonetheless, both friends and foes of embryo research raised objections to it.

Alta Charo (1995, 11) suggested that the panel's report is significantly flawed insofar as it claims to have made a determination of the moral status of the embryo: "it is impossible for a governmental body to determine the moral status of the embryo." In one sense, that is surely true. No body, government or otherwise, can determine the moral status of the embryo in the way we can, say, determine the time it will take an object dropped from a given height to reach the ground. There is no correct answer to the question, what is the moral status of the embryo?

Human beings cannot determine—in the sense of *discover* through simple empirical investigation—what the moral status of embryos is.

In another sense, however, government bodies cannot avoid determining the moral status of embryos. They have to do so in the sense of *implicitly or explicitly interpreting* moral status. What we think is appropriate to do with things is to a large extent a function of what we think the things are. When an advisory body makes a policy concerning the disposition of embryos, it has to rely on an interpretation of—it has to make a determination about—their moral status. No matter how keenly such a body might be aware that the interpretation it relies on is tentative and potentially divisive, it cannot avoid choosing an interpretation. Thus, pace Charo, I would suggest that HERP should *not* have "abandoned any effort to determine the moral status of the embryo"(Charo 1995, 18). The panel should not have attempted to avoid making a determination because no such attempt could succeed. Indeed, whereas it technically may have been correct to assert that it "was not called upon to decide which [of the many views on the moral status of the embryo] is *correct*" (1994), it should have acknowledged more clearly that it nonetheless had to base its recommendations on its interpretation.

In asserting that it was not called upon to decide which view of moral status was correct, the panel relied on the conviction that it "conducted its deliberations in terms that were independent of a particular religious or philosophical perspective." Someone in a hurry might have missed the work that "particular" does in that sentence. The report was free of particular religious commitments in the sense that it did not appeal to a particular biblical tradition or religious authority to support its interpretation of the moral status of embryos. Similarly, it was free of particular philosophical commitments in the sense that it did not appeal to any particular philosophical school such as deontology or utilitarianism. On the other hand, the very idea of democracy has deep roots in commitments that are arguably religious and surely philosophical. Although, for example, the idea that all human beings are created equal can be given a philosophical account, its religious roots are obvious. The idea that the government should, to the extent possible, allow a plurality of life projects to flourish is rooted in fundamental philosophical commitments.

The difference between *particular* philosophical and religious commitments that are not essential to the idea of democracy and more general philosophical (and arguably religious) commitments that undergird the idea of democracy is important. The so-called pluralistic approach does not come from nowhere; it is not value free. It grows out of a commitment to and traditions of giving reasons that are accessible to all. The NBAC should specify some of the essential democratic and philosophical commitments to which it will appeal, such as the commitment to giving reasons that do not decisively depend on particular schools of religion or philosophy.

This nation's founders understood that sometimes disagreements about policy matters would be rooted in deep religious and philosophical commitments. Such disagreements have to be resolved through political process. Even if the reality of political and economic power is often otherwise, in principle, those arguments prevail that persuade the majority. In accordance with that process, the government must sometimes implement determinations that conflict with the fundamental values of some citizens. It is utopian to imagine that at all times all deep commitments will be able to flourish. As John Rawls (1993, 197) put it: "there is no social world without loss: that is, no social world that does not exclude some ways of life that realize in special ways certain fundamental values." Inevitably, some citizens will sometimes feel the pain of such exclusion. We are all obliged to notice and try to respond to that pain.[3]

The founders believed that those whose values were not embraced in a given case could take solace in understanding that the procedure that produced the result was rooted in a shared fundamental value: the value of relegating such disagreements to a public arena in which those with the most persuasive position prevail. The founders were aware that history is strewn with examples of bad arguments persuading the majority. But in a democracy the remedy is not religious fiat; it is better arguments.

Because one cannot avoid making an interpretation of the moral status of embryos, in a democracy one must instead give reasons to support one's interpretation. Proponents of the intermediate status view cannot claim to be without such an interpretation; they should be clear and open about it and should feel no need to apologize for it. They should acknowledge the pain and frustration that those holding minority views will experience. Advocates of intermediate moral status can and should point to how their interpretation acknowledges minority claims: acknowledging advocates of the embryos-are-persons view, limits are placed on the time frame in which embryos can be used for research as well as on the purposes to which they can be put; acknowledging those who hold the embryos-are-property view, much, but not all research is allowed.

It was perhaps because HERP was not sufficiently clear about how it understood the intermediate status of embryos that another objection was leveled against its report. One observer suggested that it is incoherent to say that we can both respect embryos and accept their dismemberment in the research process (Callahan 1995). That would be true if, for example, one assumed that embryos are persons and thus deserving of the same respect as persons. But if, as did HERP, one conceives of the moral status of embryos differently, respecting them differently could be altogether coherent. What we think we should do with things, or how we think we should respect them, is a function of what we think they are. For example, we think we can consistently accord cadavers the respect they are due and allow medical students under carefully circumscribed conditions to dismember them. If one accepts the middle way of interpreting the moral status of embryos, limited (but appropriate) respect for them is consistent with limited research on them.

## The Discarded-Created Distinction: Intentions of Embryo Makers

Thoughtful people suggested that an important moral difference exists between doing research on embryos originally created with the intention of using them for reproduction and doing research on embryos originally created with the intention of using them for research. The former embryos become available for research only when it is discovered they are no longer required for reproduction; only then are they discarded. The latter embryos would be created specifically for the purpose of research. According to one view, doing research on discarded embryos is far easier to justify than is doing research on specifically created embryos.

It is, of course, altogether unclear how oversight bodies would be able to discern the intentions of embryo makers. Although regulations could perhaps impose limits on the number of embryos allowed to be created for in vitro fertilization (IVF), there will always be room for creative overestimation of the need for embryos for reproduction.

Whereas members of a bioethics commission have to take into account such practical concerns, it is ethical concerns that should drive their analysis. One ethical intuition that seems to motivate the discarded-created distinction is that whereas the act of creating an embryo for reproduction is respectful in a way that is commensurate with the moral status of embryos, the act of creating an embryo for research is not. Because the first class of embryo was brought into being under moral circumstances—because the intentions of its makers were moral—research on them is deemed acceptable.[4] Because the second class of embryo was not brought into being under equally moral circumstances—the intentions of its makers were not equally respectable—research on them is deemed unacceptable. In this view, the moral status of the embryo (and thus the moral status of research on it) is a function of the intention of its maker. The problem with this intuition is that it is difficult to see what the intention of the maker of something has to do with the moral status of that thing once it has come into being. We do not think, for example, that the moral status of children is a function of their parents' intention at the time of conception. If what something *is* obliges us to treat it some ways and not others, how it came into being is usually thought to be morally irrelevant.

Another and closely related motivation for taking the discarded-created distinction seriously is the intuition that, whereas in creating embryos for reproduction scientists are helping nature along toward a natural purpose, in creating embryos for research they are not. According to this intuition, helping nature along is praiseworthy, but doing something different from what happens "naturally" is not. Intending to create embryos for the purpose of reproduction is natural, intending to create them for the purpose of research is artificial. The problem is that both reproduction and research entail the intentional creation of embryos in the highly artificial context of an IVF clinic. It is difficult to see why policy makers should give credence to the natural-artificial distinction in attempts to delineate the moral difference between doing research on the two types of embryos.

Another point at work in taking the distinction seriously may be the intuition that the good of helping an infertile couple become pregnant is a greater than the good of doing embryo research. But insofar as most of that research aims at helping many couples overcome infertility, it is difficult to see why that good is of lesser moral weight than the good of helping an individual couple. If the good of helping an individual couple conceive is great enough to justify the creation of embryos, it would seem that the good of helping many couples conceive is an equally strong justification.

Another concern is that creating embryos for the express purpose of research could make us increasingly think of them as mere means to ends rather than as ends in themselves (HERP 1994, 53). In one sense this concern seems off the mark to those who hold an intermediate status view of embryos. It is clear to holders of that view that embryos deserve respect commensurate with their intermediate moral status, but it does not seem to them that embryos are ends in themselves the way persons are. Nonetheless, these individuals, too, would be concerned if creating embryos for research led to more general degradation of the respect due to entities that, if transferred, might become human beings. Thus, the worry about instrumentalization strikes me as worthy of further reflection.

A last motivation for the created-discarded distinction may be concern that allowing creation of embryos for research will increase pressure on women to donate ova for that purpose. However, the Canadian Commission suggested that *not* allowing creation of embryos for research would increase pressure on women, whereas allowing researchers to create embryos for research would decrease pressure on women in IVF programs to donate unused eggs or zygotes (HERP 1994, 56). Although this strategy might decrease pressure on women who already have undergone IVF procedures, the question remains concerning when and where else researchers will obtain eggs to create embryos. It is entirely plausible that that perceived need will create subtle or not so subtle pressure on women to donate eggs. Thus, like the concern about instrumentalization, the concern about pressure on women is not unreasonable. Unlike the concern about instrumentalization, however, this concern might be mitigated by using nonhuman ova.

Thus the attempt to show respect for embryos based on the intentions of the makers of embryos is fraught with practical and conceptual difficulties. Indeed, not the British Human Fertilisation and Embryology Authority, Canadian Royal Commission on New Reproductive Technologies, nor U.S. Ethics Advisory Board put much stock in the distinction; all three approved the fertilization of ova for research purposes.

## IVF versus SCNT: On Different Means Used to Create Embryos

If the intentions of the maker of embryos do not make a moral difference (or at least not the sort of clear moral difference suggested by some proponents), do the means used to make embryos make a moral difference? This question arises from the observation that whereas IVF performed for the purpose of reproduction is widely accepted, SCNT performed for the same purpose has been widely rejected.

Aside from concerns about risk, rejection of reproductive cloning is based on widespread worry about the psychologic consequences of producing children with means that replicate an extant genotype rather than creating a new one. However, since here we are talking about using SCNT for research, not reproduction, worries about reproducing an extant genotype (psychologic consequences for children) are not relevant. If in general we accept the limited creation of embryos for research, and if by definition harm-to-children concerns do not apply to using SCNT to do so, do we have another reason to object to or worry about using SCNT for that purpose?

One reason might be that SCNT will significantly increase the supply of embryos and thereby decrease respect or awe before them. This overlooks two facts. First, both traditional IVF and SCNT are limited by the number of available human ova; I am not aware of a reason to think that that number is going to grow fast. Second, at this point, it is more difficult to produce embryos with SCNT than with IVF; it is not reasonable to assume that researchers will rush to do SCNT. Thus it does not seem reasonable to worry that SCNT will significantly increase the number of, and thereby decrease the respect accorded, embryos in general.

There may, however, be another more substantial worry in this context. Since embryos created by SCNT are not genetically unique,[5] and since genetic uniqueness is one of the valued properties of embryos created by IVF, embryos created by SCNT may be respected less than those created by IVF. This worry strikes me as important and worthy of further reflection. To put it in humanistic terms, it is not implausible that the more we imagine ourselves to be the masters of nature, the more we will forget our fundamental indebtedness to nature. I acknowledge that it is not obvious that such an increasingly instrumental attitude must emerge. We have been able to maintain our awe before IVF embryos created for the purpose of reproduction. I can see no prima facie reason why we cannot also maintain awe before SCNT embryos created for research to promote human well-being more generally.

### Using SCNT to Create Nonviable Hybrid Entities

President Clinton's letter requesting an NBAC report mentions Advanced Cell Technology's (ACT) attempt to use SCNT to create hybrid entities. Much is at stake in whether we call ACT's entities hybrid embryos or something else, such as "embryonic cells" (Wade 1998).

If, as ACT's CEO, Michael West, suggested, we do not know what ACT's entities are, how can we say they are not embryos? If they are not embryos (and thus not capable of implanting in a uterus), why do researchers say that they would not transfer such entities in an experiment to see if they would implant?

To most speakers of English who followed the Dolly story, if you take a sheep somatic cell and fuse it with an enucleated sheep egg, you get a sheep embryo. If for some reason that embryo is not viable, we would call it a nonviable sheep embryo. To most speakers of English, if you take a human somatic cell and fuse it to an enucleated cow egg, you get a hybrid embryo. If for some reason that hybrid embryo is not viable, most would call it a nonviable hybrid embryo.

Therefore, I would recommend that we call ACT's entities hybrid embryos. (If it becomes clear that they are not viable, we should call them nonviable hybrid embryos.) Although that requires facing hard questions about the production of embryos by SCNT for research, facing those questions is preferable to violating the obligation to engage in public conversation in terms that attend to rather than obfuscate the concerns of many citizens.

If we agree to call ACT's entities hybrid embryos, what ethical questions arise? The risk that results from mixing mitochondrial DNA from one species and nuclear DNA from another is large. Mitochondrial DNA from common chimpanzees, pigmy chimpanzees, and gorillas is compatible enough with human nuclear DNA for one of the cell's basic functions, oxidative phosphorylation, to occur; mitochondrial DNA from orangutans, New World monkeys, and lemurs, however, is *not* compatible enough (Kenyon and Mores 1997).

If the risk question were resolved, the more complicated question concerning the ethics of creating hybrid organisms would remain. Anxiety about at least some forms of hybridity rests on deep but generally indefensible intuitions: perhaps the best example is the "intuition" that "miscegenation" is criminal. Yet, it is both practically and theoretically unwise for a bioethics commission to dismiss a general worry with such a long and powerful history. It was so obvious to HERP that producing chimeras is unacceptable that they did not think it necessary to give reasons for that decision. In principle at least, it would seem that although some anxiety may harbor nothing more than ignorance, some may harbor insight. The question of hybridity deserves further exploration.

However, if concerns about hybridity are really about the production of chimeras, and if ACT only wants to use an enucleated cow egg as a way station for human nuclear DNA destined to become ES cells, concerns about hybridity would in general appear not to apply. Insofar as we are willing to place genes from one species into another to produce things such as insulin or transplantable organs, it is not easy to see on what grounds we might object to housing a nuclear genome from one species temporarily in the cytoplasm of another.

If the risk and hybridity questions were resolved, ACT's strategy would hold two large benefits. First, the technique provides a way to histocompatibility: the person who needs the transplanted tissue provides the somatic cell from which the tissue is produced. Perhaps more important, insofar as ACT's strategy does not involve human ova, it eliminates concern about creating pressure on women to donate ova.

## Concluding Thoughts

These are the issues that I believe an intellectually honest and adequate response to President Clinton's request must address, or at least acknowledge.

Before taxpayer money is spent on research involving embryos, whether human or hybrid, arguments should be made that persuade the majority of taxpayers and Congress that such research is ethically acceptable. That will not be easy, but it will not be impossible. Americans are mightily and appropriately impressed by the potential medical benefits associated with embryo research. Although HERP's position on limited embryo research did not prevail, that should not keep others from trying to make more persuasive stands.

Medical progress is a very great good, but it does not trump all others. In particular, it does not trump the good of transparent and respectful public debate. It is ultimately (if not immediately) in everyone's best interest to be as clear as possible about the facts. One of those is that ES cell research cannot be done without destroying embryos, whether they are hybrid or from a single species, viable or nonviable, created by IVF or SCNT.

Beyond the fact that the president asked for a broader analysis, it is what the public needs and deserves.

## Notes

1. On January 15, 1999, Ms. Rabb wrote in a memo to Dr. Varmus, "federally funded research that utilizes human pluripotent stem cells would not be prohibited by the HHS appropriations law prohibiting human embryo research, because such stem cells are not human embryos."

2. Alternatively, genetically altered ES cells could be fused with enucleated eggs; see Resnik et al. 1999, 80.

3. Although earlier I disagreed with Charo's "Hunting the snark" (1995), here I am indebted to that essay.

4. "For most people it is the intention to create a child that makes the creation of an embryo a moral act" (Annas et al. 1996, 1331).

5. To be more precise, the *nuclear* DNA of embryos produced by SCNT is not unique; because of mitochondrial DNA contributed by the enucleated ovum, an embryo produced by SCNT *is* genetically unique. This minor technical point

does not change the fact that many may worry about the moral significance of replicating (nuclear) genotypes, even in research.

## References

Annas, G., Caplan, A., and Elias, S. 1996. The politics of human-embryo research—Avoiding ethical gridlock. *New England Journal of Medicine* 554(20): 1329–1332, 1331.

Callahan, D. 1995. The puzzle of profound respect. *Hastings Center Report* 25(1): 39–41.

Charo, A. 1995. The hunting of the snark: The moral status of embryos, right-to-lifers, and third world women. *Stanford Law and Policy Review* 6: 11–27.

Gordon, J. W. 1999. Genetic enhancement in humans. *Science* 283: 2023–2024.

Human Ethics Research Panel. 1994. *Report of the Human Embryo Research Panel*. Washington, DC: National Institutes of Health.

Kenyon, L. and Moraes, C. T. 1997. Expanding the functional human mitochondrial DNA database by the establishment of primate xenomitochondrial cybrids. *Proceedings of the National Academy of Sciences of the USA* 94(17): 9131–9135.

Rawls, J. 1993. *Political Liberalism*. New York: Columbia University Press.

Regalado, A. 1998. The troubled hunt for the ultimate cell. *Technology Review* 101(4): 4–41.

Resnik, D. B., Steinkraus, H. B., and Langer, P. J. 1999. *Human Germline Gene Therpy: Scientific, Moral and Political Issues*. Austin, TX: R. G. Landes.

Solter, D. and Gearhart, J. 1999. Putting stem cells to work. *Science* 283: 1468–1470.

Steinbock, B. 1994. Ethical issues in human embryo research. Background paper for NIH Human Embryo Research Panel (HERP) presented Feb 3, 1994 at Bethesda Marriott, Bethesda, MD.

Statement of Administration Policy. 1998. Concerning S. 1601–Human Cloning Prohibition Act.

Thomson, J. A., Liskovitz-Eldor, J. Shapiro, S. S., Waknitz, M. A., Swiergiel, J. J. Marshall, V. S., and Jones, J. J. 1998. Embryonic stem cell lines derived from human blastocysts. *Science* 282: 1145–1147.

Wade, N. 1998. Researchers claim embryonic cell mix of human and cow. *New York Times*, November 12.

# 5

# Human Embryonic Stem Cell Research: Comments on the NBAC Report

Françoise Baylis

In September 1999 the National Bioethics Advisory Commission (NBAC) released its final report, *Ethical Issues in Human Stem Cell Research*, recommending federal funding for research on the derivation and use of human embryonic stem (hES) cells from embryos remaining after infertility treatments. Specifically, NBAC recommended that:

Research involving the derivation and use of hES cells from embryos remaining after infertility treatments should be eligible for federal funding. An exception should be made to the present statutory ban on federal funding of embryo research to permit federal agencies to fund research involving the derivation of hES cells from this source under appropriate regulations that include public oversight and review. (NBAC 1999, 70)

In discussing conditions and constraints under which the research should proceed, NBAC stipulated that consent to the research should be sought only from individuals (or couples) who have already decided to discard their embryos instead of storing them or donating them to another couple; consent should be voluntary; it should not be possible for donors to designate or restrict recipients of tissues or cell lines derived from the research; sale of human embryos should remain illegal; only the minimum number of embryos necessary to derive the requisite stem cells should be destroyed; professional standards should be developed to discourage fertility clinics from increasing the number of embryos remaining after infertility treatment that subsequently would be eligible for research; imported human embryos or embryonic cell lines should be subject to the same regulations as would apply to such materials produced in the United States; and those who use or receive hES cells should be aware of their source so that they can avoid complicity with a practice they find morally objectionable (NBAC 1999, 53).

For some, the recommendations on the derivation and use of hES cells, which limit federally funded research to embryos that remain after in vitro fertilization (IVF) and introduce safeguards to prevent inappropriate and unnecessary use of such embryos, effectively balance the "tensions between two important ethical commitments: to cure disease and to protect human life" (NBAC 1999, 66). For others, these recommendations are morally objectionable because the sanctioned research requires destruction of human embryos.

My own view is that these recommendations may be an acceptable answer to the public policy question, but they are certainly a flawed answer to the mirror ethical question. This is because important aspects of these recommendations inappropriately rest on deeply problematic use of the word "respect", uncritical acceptance of current practice in fertility clinics, and an ethically questionable decision-making process.

## A Problematic Use of the Term "Respect"

The NBAC is acutely aware of divergent perspectives on the moral status of the developing human embryo:

At one end of the spectrum of attitudes is the view that the embryo is a mere cluster of cells that has no more moral status than any other collection of human cells. From this perspective, one might conclude that there are few, if any, ethical limitations on the research uses of embryos.

At the other end of the spectrum is the view that embryos should be considered in the same moral category as children or adults. According to this view, research involving the destruction of embryos is absolutely prohibited. (NBAC 1999, 49)

The NBAC believes that it cannot settle the debate between these competing views: "it is unlikely that, by sheer force of argument, those with particularly strong beliefs on either side will be persuaded to change their opinions." It does believe, however, that it is possible to adopt an intermediate position according to which "the embryo merits respect as a form of human life, but not the same level of respect accorded persons." At first glance this position, which is consistent with earlier declarations by the Ethics Advisory Board of the (then) Department of Health, Education, and Welfare and the National Institutes of Health Human Embryo Research Panel, seems a reasonable approach to an intractable problem. We can, for example, easily imagine differences in the respect we owe our parents, neighbors, elders within our community, creative artists, talented athletes, renowned scientists, and so on, that is distinguishable from the respect these same individuals are owed *as persons*. It is significant, however, that the differences we imagine have to do with the regard in which we do (or should) hold others and the common courtesies they are due based on personal relationships, social rank, or a positive assessment of their individual merit. But in the debate about moral status, respect is not simply about esteem or etiquette; rather it has to do with dignity independent of rank and merit. To respect something is to regard it as valuable in itself, to cherish it because of what it is. Following Downie and Telfer (1969, 15), "[t]he expression 'because of what it is' suggests not only why it is valuable but also what cherishing it amounts to; to cherish a thing is to care about its essential features— those which, as we say, 'make it what it is'—and to consider important not only that it should continue to exist but also that it should flourish." This explains why, in debates about moral status at the beginning and end of life, claims about respect translate into claims about protectability, and more specifically about an absolute or presumptive right to life.

In this view, with the intermediate position adopted by NBAC, there would appear to be some equivocation with regard to the meaning of "respect." Arguably, however, there is no equivocation if respect is simply a euphemism for some kind of moral consideration as decreed by the speaker or author, in which case NBAC can consistently assert that the moral consideration due persons is in the realm of protectability and includes an inviolable or prima facie right to life, whereas the moral consideration owed human embryos (presumably because they lack special human capacities common among persons) is in the realm of accountability, and this requires only that they be handled with care.

In the final report, NBAC decreed that respect for human embryos requires that:

(1) they be destroyed only with good reason—i.e., in pursuit of a worthy goal such as, research "necessary to develop cures for life-threatening or severely debilitating diseases;"
(2) they be destroyed only for research purposes when "no less morally problematic alternatives are available for advancing the research"—i.e., the worthy goal could not be achieved by other less morally problematic means; and

(3) they not be the object of sale before their destruction (although progenitor gametes may have been purchased), as this would contribute to commodification of human embryos. (NBAC 1999, 52–53)

By stipulating these elements, NBAC seeks to rationalize the killing of embryos and at the same time to "demonstrate *respect* for all reasonable alternative points of view" (italics added). Some, no doubt, will be satisfied with this approach. Others, however, will ask the pointed question: "What in the world can that kind of respect mean?" (Callahan 1995) For certainly something is very odd in claiming to cherish the human embryo because of what it is while at the same time planning for its destruction.

## A Mistaken View of the Normal as the Moral

Current practice in fertility clinics is such that individuals (or couples) who consent to the creation of embryos can discard these embryos if they are either unsuitable (e.g., nonviable, or afflicted with a serious genetic disorder) or no longer required for treatment. Commenting on this practice, NBAC (1999, 52–53, 66) noted that embryos about to be discarded after infertility treatment "have no prospect for survival even if they are not used in deriving ES cells ... the research use of such embryos affects only how, not whether, the destruction occurs ... it [stem cell research] does not alter their final disposition." On this basis, "[i]f embryo destruction is permissible, then it certainly should be permissible to destroy them in a way that would generate stem cells for bona fide research" (NBAC 1999, 53). This conclusion appears eminently sensible; it is nonetheless problematic. The conclusion is a conditional statement (if ..., then ...), and the antecedent "if" clause is neither obviously true nor argued for. The permissibility of destroying human embryos remaining after infertility treatment is a morally contested issue, and thus so too is the permissibility of destroying them in a way that would generate stem cells.

In response to this criticism, it might be maintained that the concluding statement is loosely phrased simply as a heuristic device to emphasize the permissibility of destroying unwanted embryos in such a way as to produce stem cells for research. If so, this statement involves a kind of epistemic mistake—inferring the acceptability of a new or contested practice (i.e., stem cell research on embryos remaining after infertility treatment) based on its resemblance to an existing practice that appears to be widely accepted (i.e., destroying embryos remaining after infertility treatment) without critically examining the defensibility of the existing practice.[1] Destroying embryos is a common practice in fertility clinics. It does not follow, however, that this is permissible except insofar as it is not legally actionable. The frequency of an activity does not in itself establish its moral permissibility. Slavery was once common, but never morally permissible.

As a rejoinder, attention may then shift to the permissibility of abortion, which involves the destruction of the fetus, a human being at a later stage of development than the embryo. The legitimacy of destroying embryos remaining after infertility treatment, however, cannot be inferred from the fact that women have the right to abortion, as the two situations are not analogous. With abortion a conflict exists between the interests of the pregnant woman and the fetus. In marked contrast, whereas "the existence of the fetus may directly conflict with the pregnant woman's interests, ... a particular ex utero embryo does not threaten anyone's interests" (NBAC 1999, 52). Furthermore, with abortion it is possible to draw a distinction between a right to evacuate and a right to destroy, and to contend that whereas the woman has a right not to be pregnant, she does not have a right to kill her fetus per se, and so clearly would not have a corresponding right to kill her embryo. Abortion is permissible only because of physical limitations of the situation and limited available technology, which make it such that only one set of competing interests can be satisfied. With destruction of unwanted human embryos not only are there no competing interests, there are also not the physical and technological limitations that are relevant to the abortion debate. And from a legislative perspective, it is important to remember that the constitutional protection of early-term abortion (*Roe v. Wade* 1973), rests on the right to privacy, and the right of a woman to control her own body, rather than the rights of scientists to use a living embryo for other purposes (Healy and Berner 1995).

The point is that NBAC does not provide a rationale for the moral permissibility of destroying unwanted viable human embryos for no reason other than they are no longer needed by those for whom they were created or to whom they were donated. This is a problem insofar as the argument for the moral permissibility of destroying embryos for research purposes builds on the presumption that discarding unwanted viable embryos is morally permissible. Contrary to current practice, however, perhaps individuals (or couples) who consent to the creation of embryos for reproductive purposes should not have the option of discarding unwanted viable embryos. Perhaps all unwanted viable embryos should be donated to others for infertility treatment; and if this is not acceptable, then at the outset, individuals (or couples) should limit the number of embryos created to those they are willing, in principle, to have transferred in one or more cycles. In any case, the burden of persuasion in establishing the moral permissibility of destroying viable human embryos remaining after infertility treatment rests with NBAC. If that burden cannot be met, the prudent course of action would be to err on the side of caution and not to destroy viable embryos either by discarding them or using them for research purposes.

## A Flawed Decision-Making Process

Having restricted the source of material for hES cell research to embryos originally created for reproductive purposes, NBAC stipulated that consent to research use of these embryos must be sought only after it has been decided that they are to be discarded.

Prospective donors of embryos remaining after infertility treatments should receive timely, relevant, and appropriate information to make informed and voluntary choices regarding disposition of the embryos. Prior to considering the potential research use of the embryos, a prospective donor should have been presented with the option of storing the embryos, donating them to another woman, or discarding them. If a prospective donor chooses to discard embryos remaining after infertility treatment, the option of donating to research may then be presented. (NBAC 1999, 72)

Two reasons are given for the requirement that the decision to discard unwanted human embryos precede the decision to donate them for research purposes: "concerns about coercion and exploitation of potential donors, as well as controversy regarding the moral status of embryos" (NBAC 1999, 72). As regards the latter issue, NBAC hopes to side-step the debate about the moral status of human embryos; stem cell research deemed eligible for federal funding only determines the manner in which the embryos will be destroyed, not whether they will be destroyed. On the other hand, concern about coercion and exploitation is more difficult to understand.

It is not clear to me, for example, how an individual (or couple) given the option to store embryos, donate them to another woman, or discard them is at increased risk of coercion or exploitation if told, at the outset, that there are two ways of discarding embryos, one of which involves research. To be sure, persons who avail themselves of the new reproductive technologies are at risk of coercion and exploitation. It does not seem reasonable, however, to try to address this problem by withholding (temporarily) information germane to their decision making regarding the disposition of unwanted embryos (assuming, for the sake of argument, that this practice is morally permissible). Imagine the following: a couple is presented with the option to store, donate, or discard their unwanted embryos. They have no desire to store the embryos as their reproductive goals have been achieved. They have concerns about donating their embryos to another individual (or couple) because they are uncomfortable with the idea that another family would raise "their" child(ren). Discarding the embryos seems unacceptably wasteful, however, and so despite their misgivings they agree to donate their unwanted embryos to another infertile individual (or couple). They never learn of the option of donating the embryos for research purposes, an option they would have chosen had they been provided with full information. Arguably this outcome is the result of a subtle form of coercion. If the fertility clinic has an embryo-donation program but is not actively involved in embryo research, this outcome might also be construed as exploitative.

To be fair, this hypothetical example is in many respects unrealistic. Informed choice is an iterative and interactive process, and the couple's concerns presumably would have been known and addressed before a final decision was made. Second, with media attention on embryo research and the typical information provided by fertility clinics, fe if any couples (or individuals) in an IVF program could be completely unaware of the option of donating embryos for research. In particular, NBAC recognized that the proposed separation between the decision to discard and the decision to donate for research "may not be possible, ... because the couple may be given several options simultaneously, either at the outset of treatment for infertility or after its completion ... But even then, it may be appropriate to view the options as consisting of donation of the embryos to another couple, their continued storage, or their destruction, with destruction of the embryos taking one of two forms" (NBAC 1999, 53). The problem with this proposal, however, is that although discarding embryos and donating them for research both result in destroying the embryos, these are not equivalent options and, as illustrated, they may not hold the same place in any rank ordering of preferences. For this reason, among others, it is strange that NBAC would advocate temporarily withholding relevant and appropriate information, particularly as this would appear to contradict its own requirement for the timely disclosure of information.' More puzzling still is that it would advocate a practice that undermines substantial understanding, a key element of the informed consent process as discussed by one of commissioners in his earlier writings (Beauchamp and Childress 1994).

The only way I can make sense of the requirement that the decision to destroy unwanted embryos precede the decision to donate them for research is to interpret this as a purposeful effort to align the new practice of donating embryos for research with the accepted practice (and existing laws and policies) of donating fetal tissue for research. The reason for so doing presumably would be strategic, namely, to limit possible objections to embryo research. But the strategy has serious limitations, not the least of which is that it leads NBAC to propose a flawed decision-making process.

## Conclusion

In response to these diverse challenges, some may hold that NBAC's final set of recommendations for research on the derivation and use of hES cells is as good as it gets, given the constraints of political reality, which include having to recommend "justifiable policy for a secular government in a country of wide-ranging religious beliefs" (Charo 1995, 607). In this view, one cannot expect more from a national commission in a pluralistic democracy with a mandate to recommend a public policy that government can sell to persons in the policy and political process, including the public (Brock 1987). Perhaps not. But what does this say about us and about democracy in the twenty-first century? More precisely, what does this say about the value we place on careful ethical reflection, the prospects we have for meaningful transparency in the policy-making process, and the confidence we have (or rather lack) in the possibility of serious and sustained public debate among concerned citizens with regard to sensitive and complex ethical issues that the future inevitably holds?

## Acknowledgment

Research for this paper was supported by a research leave grant from Dalhousie University and a bioethics grant-in-aid from Associated Medical Services Inc.

## Notes

1. According to Schrecker et al. (1998, 168), this type of epistemic mistake may be of particular importance in debates about the use of biotechnology. Consider, for example, the proposal that germ line enhancement technology is permissible "because society already permits, and even encourages, parents to 'improve' their children in various ways."

2. The NBAC (1999, 72) recommended disclosure in stages except in cases in which prospective donors ask astute questions that warrant full disclosure. "At any point, the prospective donors' questions—including inquiries about possible research use of any embryos remaining after infertility treatment—should be answered truthfully, with all information that is relevant to the questions presented."

## References

Beauchamp, T. L. and Childress, J. F. 1994. *Principles of Biomedical Ethics*, 4th ed. New York: Oxford University Press, 1994.

Brock, D. W. 1987. Truth or consequences: The role of philosophers in policy-making. *Ethics* 97: 789.

Callahan, D. 1995. The puzzle of profound respect. *Hastings Center Report* 25(1): 39–40.

Charo, A. 1995. Embryo research: An argument for federal funding. *Journal of Women's Health* 4(6): 607.

Downie, R. S. and Telfer, E. 1969. *Respect for Persons.* London: George Allen & Unwin.

Healy, B. and Berner, L. S. 1995. A position against federal funding for human embryo research: Words of caution for women, for science, and for society. *Journal of Women's Health* 4(6): 609–613.

National Bioethics Advisory Commission. 1999. *Ethical Issues in Human Stem Cell Research.* Vol. 1. *Report and Recommendations of the National Bioethics Advisory Commission.* Rockville, MD: National Bioethics Advisory Commission.

*Roe v. Wade.* 1973. No. 70–18, 410 U.S. 113, 93 S. Ct. 705.

Schrecker, T., Hoffmaster, B., Sommerville, M., and Wellington, A. 1998. Biotechnology, ethics and government: Report to the Interdepartmental Working Group on Ethics. In: *Renewal of the Canadian Biotechnology Strategy Roundtable Consultation: Background Documents.* Ottawa: Industry Canada.

# 6

# NBAC's Arguments on Embryo Research: Strengths and Weaknesses

John C. Fletcher

Stem cell research not only galvanized popular opinion, it immediately generated a series of position papers from national public commissions. This chapter will discuss the ethical arguments behind one such report by the National Bioethics Advisory Commission (NBAC), and offer what I think would be a stronger argument for future policy. The NBAC was asked to address the moral questions raised by the derivation of human embryonic stem (hES) cells from excess embryos regardless of the source of funding, and to consider whether it is good public policy to use federal funds for this purpose (Fletcher 2000). The report stated that hES cell research is justified because of the potential benefits for healing and prevention of disability. But different bodies, such as the National Institutes of Health (NIH) and American Academy for the Advancement of Science (AAAS) made different judgments about whether uses of hES cells ought to be viewed differently, in a moral sense, than derivation of the cells from embryos that are destroyed in the process. Furthermore, they differed on whether a distinction between derivation and use was of importance in public policy. What ethical and practical judgments guided these similar and different conclusions?

## hES Cell Research: Why Not Take the Least Offensive Moral Approach?

Fetal tissue research is morally acceptable to many because it can be compared with cadaveric organ retrieval after homicide or suicide. Furthermore, in this case, the woman's voluntary choice to donate is protected by informed consent, legal safeguards exist to prevent abuses, and research offers an opportunity to learn to heal and prevent disabilities. The case of hES cell research with excess embryos is similar in every respect except that researchers' actions cause embryos to die; in fetal tissue research the abortion has already occurred and the fetus is dead. This dissimilarity poses an inescapable question: can it be morally right to destroy embryos?

What ethical perspective does NBAC bring to this question? The authors framed its moral position largely in terms of loyalty to medicine's goals of healing, prevention, and *research* (italics mine). They envisioned these goals as "rightly characterized by the principles of beneficence and nonmaleficence" (NBAC 1999, 69). The principles approach would frame the moral problem of deriving hES cells from embryos as a conflict between obligations shaped by the principles of beneficence and nonmaleficence. Considerations of beneficence encourage research aimed at healing and preventing of disability, but only when conducted under existing or heightened moral guidelines. Judgments drawn from nonmaleficence lead to restraints on research in the form of safeguards against harms and abuses, which NBAC carefully specified. Obligations of respect and protection for developing embryonic and fetal life are also grounded in the principle to avoid harm and prevent it wherever possible. However, the obligations of beneficence are given more weight than all the claims that flow from loyalty to avoiding harms, especially harms to embryos. Given its recommended safeguards and the anticipation of "great" therapeutic benefits, NBAC recommended that Congress ought to amend the ban to permit federal funding of research involving excess embryos.

Could the authors' moral premise be flawed? They assert, with no argument, that research is a goal of medicine equal to healing and prevention. Let me propose that elevation of research is a serious error.[1] Research that is ethically controlled certainly serves the goals of medicine but is not an independent goal of medicine. In the words of Hans Jonas (1969), elevating biomedical research so high in the heirarchy of values would pose dangers "... by the erosion of those moral values whose loss, caused by too ruthless a pursuit of scientific progress would make its most dazzling triumphs not worth having." Jonas understood research to be subservient to ethical considerations of medicine that were dependent on larger judgments of society, which he called moral values. His understanding of the subordinate moral status of research is the prevailing understanding in our society. Many will have to be convinced that hES cell research is not "too ruthless" a pursuit of healing and prevention that will make even the "dazzling triumph" of cell-replacement therapy not worth having. The NBAC's loyalty to the cause of research prevented openness to a compromise position on federal funding that would reassure conservatives that their concerns had truly been weighed in the balance that NBAC aimed for but failed to achieve.

The NBAC is persuasive in its search for common ground between liberals and conservatives on the morality of hES cell research. The reasoning closely follows Andrew Siegel's commissioned paper (1999) that employs the resources of casuistry and again offers the case of abortion as a moral analogy to embryo research. Siegel[2] built on insights from Dworkin's (1993) important work on abortion and euthanasia, which is cited in the NBAC report. Dworkin noted that many conservatives are not moral absolutists on abortion, despite their extreme rhetoric. They make exceptions for abortion in cases in which the pregnancy endangers the woman's life, or rape or incest has caused the pregnancy. In these cases, they are willing to give priority to the interests of other existing persons over those of the fetus. If conservatives are morally consistent, NBAC argues per Dworkin, it is reasonable that some may willingly cross the gap to permit hES research to save lives or prevent disability, especially if strong safeguards are in place.

Are the two cases morally more similar than dissimilar? Will the exceptions that conservatives grant in some cases of abortion hold up when the case is deriving hES cells from embryos? The NBAC acknowledges that abortion and embryo research are different in one important moral feature: an unwanted fetus collides with a woman's interests, but an embryo's existence threatens no one. However, they contend that "the two cases share an implicit attribution of greater value to the interests of children and adults ..." than to the interests of the fetus or embryo. In short, if one would be willing to concede that abortions are morally acceptable in cases of rape, incest, and to save a woman's life, one ought to be willing to concede that destroying embryos is morally acceptable to save lives and prevent disability.

But is hES cell research really a similar case? The results of the research are distant and uncertain by comparison with the immediacy of saving a woman's life by abortion. The NBAC's own scientific chapter states: "Research into the use of EG [embryonic germ] and hES cells is still at an early stage" (NBAC 1999, 20).

The authors wrote:

Research that involves the destruction of embryos remaining after infertility treatments is permissible when there is good reason to believe that this destruction is necessary to develop *cures* [italics mine] for life-threatening or severely debilitating diseases and when appropriate protections and oversight are in place ... to prevent abuse. (NBAC 1999, 52)

This statement links hES cell research too confidently with cures. The science at that time did not justify this much assurance. *Even today such a statement would be premature.* What is really going on here is exaggeration, as exemplified by the added statement "great promise." Why the inflated language? It is fanfare for a large U-turn in the argument, which has a stated and an unstated aim. The stated aim is to find common ground in the morality of hES cell research that conservatives can share. The unstated aim is to strengthen the case for federal support of the research, which most of the NBAC desired. At this crucial point, the unstated aim starts to drive the argument. The search for "shared views" is over. The authors had to fit together a moral thesis to justify destroying embryos in research and a recommendation that Congress amend the ban *now*. An appeal to consequences citing great therapeutic promise generates urgency for both goals, but any appeal to consequences cries out for scientific evidence.

Here, the NBAC states:

Some might object ... that the benefits of the research are too uncertain to justify a comparison with the conditions under which one might make an exception to permit abortion. But the lower probability of benefits from research uses of embryos is balanced by a much higher ratio of potential lives saved relative to embryonic lives lost and by two other characteristics of the embryos used to derive hES cells: first, that they are at a much earlier stage of development than is usually true of aborted fetuses, and second, that they are about to be discarded after infertility treatment and thus have no prospect of survival even if they are not used in deriving hES cells.

The authors claim that the imbalance "is corrected" by a "much higher ratio" of potential lives saved than embryonic lives lost, plus two characteristics of embryos. Hence they conclude that the "potential benefits of the research outweigh the harms to embryos that are destroyed in the research process" (NBAC 1999, 52).

What scientists hope will be therapeutic benefits of hES cell research, however, is not the same as data in animal experiments from which to plot the way to the threshold of that goal with confidence. The scientific chapter had it right: "research into the use of EG and hES cells is still at an early stage"(NBAC 1999, 20). Current scientific study is focused on the capacity of hES cells to continue growing and dividing. Some adult stem (AS) cells seem to lose this ability (Vogel 2000). These differences raise hopes for the potential of hES cells to outlive AS cells and be more effective in the long run for therapy. However, impressive results are being obtained by preclinical research in mice with AS and not hES cells. The jury is still out on whether animal research will find that hES cells are the royal road to therapy, one road among several, or a road unsafe to take.

Without data, NBAC's "ratio" is speculation and not substance. The reader is supposed to take it on faith, surely the faith of the liberal scientists and the majority of NBAC at the time, that the ratio was favorable. Conservatives will not be drawn to this faith, and neither is this pragmatic reader, but for very different reasons.

To increase weight to the ratio of potential lives saved relative to embryonic lives lost, the authors add on two "characteristics" of embryos to be used: embryonic life is an earlier stage of development than fetal life and these embryos will be otherwise discarded. But underneath these "facts" are appeals to values. Indirectly and without comment, the authors ask the reader to accept the proposition that embryos are of less value than fetuses, and furthermore, that to discard rather than conduct research with embryos is a loss of opportunity for benefits; that is, a loss of value. Many will be insulted by the implication that they could assent to these implicit values, since they hold that the moral value of human life is intrinsic and independent of any stage of development. But the unguarded assertion that derivation can be justified because embryos will be discarded is based on the theory that it can be morally right to kill a doomed human being to benefit others.

A second objection to hES cell research is that there may well be morally preferable alternatives to it. In his written testimony before NBAC on April 16, 1999, Doerflinger noted that AS cell research is making hES cells irrelevant. The NBAC agreed that "the derivation of stem cells from embryos remaining following infertility treatments is justifiable only if no less morally problematic alternatives are available for advancing the research." The NBAC report objected on scientific grounds.

They assumed, on the basis of scientists' testimony, that hES cells are more promising for therapy than AS cells. They thus placed the burden of scientific proof on those who believe that AS cells are a preferable alternative. But all cell lines raise unsolved scientific questions, for example, hES cells inserted into mice are tumorigenic (Solter and Gearhart 1999). Can cell lines grown from hES cells be cleansed of this danger? Since NBAC's report was written, one report regarding EG cells (Steghaus-Kovac 1999) posed serious questions about safety their for human cell transplants.[3]

In my view, the main strength of the moral argument was the genuine point of contact with conservatives' willingness to make exceptions for some abortions, but it did not complete the moral reasoning required by this analogy. In U.S. public policy, these exceptions are incorporated into the Medicaid program. But no other exceptions are acceptable for federal funding. The weakness of the position, I believe, is failure to see that a similar, practical compromise might be employed for this case—an error made in the service of eagerness to fund hES research.

## NBAC's Recommendations on Federal Funding

The report proposes federal support of EG cell research with fetal tissue and hES cell research with excess embryos. It opposes federal support with embryos created only for research or with somatic cell nuclear transfer embryos. Not lost on NBAC was opposition to the NIH panel's recommendation of federal support for research with embryos created only for this purpose (Marshall 1994; Callahan 1995; Campbell 1995; Annas et al. 1996), as well as the partial dissent of Patricia King (1994).

The NBAC gave eight reasons why Congress should exempt hES research from the ban on fetal research: scientific reasons justify it, that is, discoveries are possible by funding derivation of hES cells; properties of hES cells depend on conditions for deriving them; and instability of hES cells in culture requires rederivation. As well, public-private synergy will shorten time to therapeutic trials, and NIH's legal position (no funding for derivation, funding for downstream research) is morally flawed and diminishes the scientific value of the whole enterprise of hES cell research. In addition, federal regulation of derivation from discarded embryos will prevent use of these embryos for hES cell research, and relying only on cell lines derived by private industry limits competition, scientific progress, and dissemination of results. They also recommended a new federal oversight body for pluripotential stem cell (PSC) research to make federal funding available and ensure accountability and public scrutiny.

These reasons are plausible in themselves; however, the issue is the politics of embryo research. Significant political voices, beginning with the Clinton administration, disagreed with NBAC's position on amending the ban. The NBAC concluded its public deliberations on federal funding on July 14, 1999 (Weiss 1999). The same day, the White House issued a press release praising NBAC's "hard work" on "a complex and sensitive matter" but also stating, "No other legal actions are necessary at this time, because it appears that human embryonic stem cells will be available from the private sector." Before NBAC's release of its report, the AAAS-ICS report agreed with the White House and NIH that the ban should remain in effect (Chapman et al. 1999). Pending some dramatic breakthroughs, amending the ban will not be politically feasible in the immediate future. Another factor at the time of the NBAC vote was that, in the midst of congressional debate on a record $17.9 billion NIH budget, neither the White House nor the NIH wanted confrontation on hES cell research.

The political situation in Congress is volatile. Twenty senators, including John McCain (Wadman 2000), signed a letter of opposition. However, some influential conservatives such as Senator. Thurmond (R-S.C.) and Senator Specter (R-Pa.) support the NIH position (MacIlwain 1999). Senators Specter and Harkin (D-S.D.) introduced a bill to amend the ban to permit NIH funding of derivation of hES cells from excess embryos (Vogel 2000).

The political stakes are high. The numbers of Americans with cell-wasting diseases—and their caregivers—vastly outnumber those who would, on principle, forgo the potential benefits of hES cell research to protect discarded embryos (Perry 2000). Large numbers of voters will be watching the political process. In my view, a moderate position to leave the ban unchanged and permit NIH downstream funding will offend the fewest of these voters.

Yet if this form of treatment turns out to be safe and effective, a difference of five or ten years will affect millions of people and their families (Perry 2000). Funding by NIH and National Science Foundation (NSF) of both derivation and use of hES cells will probably speed progress, as desired by NBAC. The rate of progress to trials of cell-replacement therapy is a political and ethical issue. Specifically, rate of progress to trials raises issues of distributive justice.

But the rate may well be slower than suggested by the enthusiasm of NBAC. Many types of experiments will precede clinical trials of cell-replacement therapy. For example, Brigid Hogan (1999) noted differences in DNA modification between mouse EG and hES cells. The scientific and (possible) clinical import of these differences needs exploration. John Gearhart also outlined basic research questions, such as ways to assay blastocysts for their potential of yielding hES cells (perhaps by searching for genes that predispose for this capacity) to produce more cell lines than the five grown by Dr. Thomson's work, as well as other intrinsic or extrinsic factors that foster success. Another task is to study differences between cell lines derived from hES and EG cells and those grown from AS cells. Research partially relevant to this question was cited above (Steghaus-Kovac 1999).

It will be mandatory, before human trials, to show that purified cell lines derived from hES cells or other sources are not tumorigenic in mice or other animals. Research will have to repeat the one dramatic experiment using hES-derived nerve cells that restored partial spinal cord function in paralyzed rats (McDonald et al. 1999; Wickelgren 1999). Clinical trials will require careful study. The AS cells may appear to be a morally preferable alternative to hES cells. It may turn out that AS, EG, and hES cells have different potential to treat different diseases.

These studies will aim to answer this question: is cell-replacement therapy safe and effective in human beings?[4] A phase I clinical trial has already been done with mesenchymal stem cells (Horowitz 1999) for allogenic bone marrow transplantation in three children with osteogenesis imperfecta, with increases in new bone growth and prevention of fractures.

We then turn to issues of distributive justice, which bear directly on two issues in stem cell research. The first is appropriation of funds to hasten the move to clinical trials. Is it fair to the more than one million Americans (Perry 2000) who suffer from illnesses that might be treated by cell replacement, and who are also taxpayers, to maintain the ban and delay progress to possible human trials? Federal support could mean transition to trials of some four to six years rather than eight to ten years, together with the possibility of reducing deaths and disabilities among affected and at-risk persons. The ethical question is whether any delay, even to gain evidence to justify amending the ban, is justifiable.

Congress could vote to amend the ban with assurance that "no less morally problematic alternatives are available for advancing the research" (NBAC 1999, 53). This way gives the least offense to conservative moral views and is morally balanced by the compromise of the principle of beneficence required of liberals, who see the ban in "conflict with several of the ethical goals of medicine, especially healing, prevention, and research" (NBAC 1999, 69).

## Federal Funding of hES Cell Research for Transition to Clinical Trials

Let me summarize the position that NBAC could take: what would be the conditions that would allow for federal funding with a fuller national consensus?

First, that the NIH-NSF downstream peer review and funding of hES and EG cell research combined with support from the private sector has led to understanding of cell differentiation, differences among hES, EG, and AS cells, and other scientific goals of basic research. Next that scientists have successfully conducted experiments in animal models with hES cells and have established that tumorigenic dangers and other potential hazards[5] can be avoided. Next that the NIH-Food and Drug Administration-NSF scientific peer review process agrees that sufficient preclinical scientific data exist to move to clinical trials with cell lines to be grown from hES cells in one or more diseases that are life threatening or severely debilitating, because of the particular promise of hES cells in treating those diseases.

Next that scientific experts make the case that federal funding of derivation of hES cells is required as a last resort to grow cell lines for such trials, because other sources of PSCs will probably not be effective.

We would have to ensure that appropriations may only be used to fund PSC research in centers that ensure that institutional review board approval has been obtained for a two-stage consent process that separates in vitro fertilization decisions from decisions to donate embryos for research and the protection of the privacy of donors, that such research conforms to guidelines recommended by NBAC and NIH, and that fairness in selection of subjects to donate excess embryos will be assured.

The obligations of justice are critical in selecting both donors of embryos and participants in clinical trials of cell-replacement therapy. The Belmont report states that claims of justice in research activities require fair distribution of benefits and burdens of such activities over a whole population.[6] Federal noninvolvement in infertility research and the ban on federal funding for embryo research have combined to infringe on obligations created by this principle in one actual and one future way. First, the composition of the pool of donors of embryos is limited to private patients in infertility centers. Second, if not prevented by a deliberate plan, selection of subjects to participate in clinical trials of cell-replacement therapies could be biased by inequities that inhibit access to these trials, especially for poor and disadvantaged Americans. This sort of balanced policy would be the model for research on topics of significant public debate, pragmatic, yet sensitive to competing moral appeals.

## Conclusion

I believe that NBAC's work is seriously unbalanced because it did not allow the moral logic of compassionate compromise to guide its choices and recommendations on federal funding. Public bioethics is unavoidably political because it aims to influence public policy. However, the authors neglected significant factors in the political context in building their case for ES cell research and federal funding. Much more attention should have been given to the history of congressional actions on fetal and embryo research, the political weight of diverse moral views, and a volatile political context. I believe that public bioethics must be concerned with the politically possible to achieve the right balance between competing factors, especially on such a controversial issue.

## Notes

1. I made the same error in drafts of my commissioned paper (Fletcher 2000) but corrected it in the final draft after reflecting on Jonas (1969) and Miller and Brody (1995). The authors' inclusion of research as a goal of medicine may be a legacy from my earlier mistake, but the authors would surely accept responsibility for their own work.

2. Siegel was a member of the NBAC staff at the time that he wrote the paper.

3. Work in the United Kingdom and Germany indicated that when hES cells were injected into mouse embryos, tissues appeared to develop normally. However, when EG cells were injected, skeletal deformations and oversized fetuses resulted. The problem may be that EG cells, which are derived at a later time in development, negate genomic imprinting in the mouse embryo and create abnormalities of body size.

4. The Food and Drug Administration (FDA) may also play a role in this stage by helping sponsors to determine proactively what sorts of pharmacology-toxicology studies have to be done to allow first entry into humans. It is expected that, in concert with formal FDA regulatory requirements, there will be additional requirements for public and societal discussion as is currently done by the NIH/ORDA/RAC and FDA for gene therapy, and by the FDA/CDC/ NIH/HRSA public forum for xenotransplantation.

5. Some of these potential hazards are genetic abnormality, aberrant developmental processes, and functional aberrations (e.g., inappropriate or uncontrolled insulin secretion).

6. The FDA and NIH have increasingly mandated equity in access to clinical trials, notably, to require gender equality and (in the case of the FDA) to redefine pediatric labeling and study requirements. The FDA and NIH could be meaningful copartners in applying access requirements equally to publicly and privately funded studies.

## References

Annas, G. J., Caplan, A., and Elias, S. 1996. The politics of human embryo research—Avoiding ethical gridlock. *New England Journal of Medicine* 334: 1329–1332.

Callahan, D. 1995. The puzzle of profound respect. *Hastings Center Report* 25(1): 39–40.

Campbell, C. S. 1995. Awe diminished. *Hastings Center Report* (Jan–Feb): 44–46.

Chapman, A. R., Frankel, M. S., and Garfinkle, M. S. 1999. Stem cell research and applications. Monitoring the frontiers of biomedical research. Prepared by

72    *John C. Fletcher*

the American Association for the Advancement of Science and Institute for Civil Society, Washington, D.C.

Dworkin, R. 1993. *Life's Dominion.* New York: Knopf.

Fletcher, J. C. 2000. *Deliberating Incrementally on Pluripotential Stem Cell Research.* Vol. II, *Commissioned Papers for the National Bioethics Advisory Commission.* Rockville, MD. E1–E50.

Jonas, H. 1969. Philosophical reflections on experimenting with human subjects. *Daedalus* 98: 245.

Hogan, B. L. M. 1999. Statement to NBAC, Feb. 3, p. 3.

Horowitz, E. M., Prokop, D. J., and Fitzpatrick, L. A. 1999. Transplantability and therapeutic effects of bone marrow derived mesenchymal cells in children with osteogenesis imperfecta. *Nature Medicine* 5: 309–313.

King, P. 1994. *Report of the Human Embryo Research Panel.* Vol. 1. Rockville, MD: A3–4.

McDonald, J. W., Liu, X. Z., Qu, Y., Liu, S., Mickey, S. K., Turetsky, D., Gottlieb, D. I., and Choi, D. W. 1999. Transplanted embryonic stem cells survive, differentiate, and promote recovery in injured rat spinal cord. *Nature Medicine* 12: 1410–1412.

Marshall, E. 1994. Human embryo research: Clinton rules out some studies. *Science* 266: 1634–1635.

Miller, F. G., and Brody, H. 1995. Professional integrity and physician-assisted death. *Hastings Center Report* 25: 8–17.

National Bioethics Advisory Commission. 1999. *Ethical Issues in Human Stem Cell Research.* Vol. I. *Report and Recommendations of the National Bioethics Advisory Commission.* Rockville, MD: National Bioethics Advisory Commission.

Perry, D. P. 2000. Patients' voices: The powerful sound in the stem cell debate. *Science* 287: 1423.

Siegel, A. W. 1999. *Locating Convergence: Ethics, Public Policy and Human Stem Cell Research.* Vol. II. *Commissioned Papers for the National Bioethics Advisory Commission.* Rockville, MD: J1–11.

Solter, D. and Gearhart, J. 1999. Putting stem cells to work. *Science* 282: 1468.

Steghaus-Kovac, S. 1999. Ethical loophole closing up for stem cell researchers. *Science* 286: 31.

Vogel, G. 2000. Can old cells learn new tricks? *Science* 287: 1418–1419.

Wadman, M. 2000. Republican candidates clash on fetal tissue in research. *Nature* 303: 694.

Weiss, R. 1999. Panel backs easing ban on embryo research. *Washington Post*, July 15, p. A-3.

Wickelgren, I. 1999. Rat spinal cord function partially restored. *Science* 286: 1826–1827.

# 7

# Beyond the Embryo: A Feminist Appraisal of the Embryonic Stem Cell Debate

Suzanne Holland

... research should be evaluated not only in terms of its effects on the subjects of the experiment but also in terms of its connection with existing patterns of oppression and domination in society.
(Sherwin 1992, 174–175)

The debate over the ethics of human embryonic stem (hES) cell research must be placed in the larger context of which it is a part. As Sherwin's quotation suggests, this is in the context of existing patterns of oppression and domination in society, particularly for women and the poor. To date, much of the ethical concern about embryonic stem cells has focused on the moral status of the embryo and the question of whether deriving human stem cell lines from the embryo jeopardizes its moral status, its claim to potential personhood. But perhaps we should also ask whether the personhood of women and people on the margins is at stake as well. Whereas much has been said about the embryo, comparatively little has been said about the effects of stem cell research on women and the poor in the context of the larger system of health care access and resource allocation. Unfortunately, it is too easy to separate the embryo from the woman in terms of our policy decisions, as the history of the struggle for reproductive freedom has shown us.

So, the question is: in the realm of embryonic stem cell research and regenerative medicine, can we construct public policy that adequately accounts for the full personhood of those on the margins, especially women of color and working-class women? Does the National Bioethics Advisory Commission (NBAC) address these issues? The NBAC report (1999) is a first step in this regard, but it does not go far enough and it is hampered, in my view, by overreliance on concerns about the embryo so that it skirts some concrete ethical issues for the marginalized.

My concern is with justice; specifically, that women—particularly poor women, women of color, and their children—are dealt a fair hand with respect to the uses and social costs of genetic technologies in general and stem cell technology in particular.[1] Since, as I believe, late advanced capitalism has dealt such women a bad hand, a hermeneutic of suspicion ought rightly to be applied to the whole nexus in which researchers, funders, ethics boards, and corporations work. As feminist historians of science and feminist bioethicists show, research is never neutral and does not occur in a vacuum; it reflects values and commitments. Similarly, embryonic stem cell research conducted in the private sector has particular implications for particular kinds of persons and can be seen to be connected to existing patterns of domination and oppression in society about which we ought to be suspect.

In short, I am concerned with the fact that the debate over ethics of embryonic stem cell research has not squarely faced the needs and concerns of women, especially poor women. The health care needs of these women should be brought to the table lest our policy formulations once again proceed without those voices, yielding a policy that may be strategic, but is not ethically satisfying.

**Feminist Ethics: A Framework**

I begin by examining what is meant by a commitment to the full personhood of women and other marginals, using the experience of women as paradigmatic of what Iris Young (1990) called "scaled and weighted bodies" in our culture, bodies on the margins. I do not claim that women have some monolithic common experience, of course, only that the historical pattern of their treatment in general reflects an attitude of disregard for other not-male, not-white persons. I offer the following basepoints for assessing a social and ethical commitment to full personhood for these individuals.

It seems to me that, at a minimum, an ethical commitment would entail social policy that reflects a fundamental trust in the moral agency of women and those on the margins, which is to say recognition that all persons deserve the opportunity to make legitimate choices about conditions that affect their lives, and are deserving of respect whenever they exercise such agency. Second, I suggest that social policy reflecting commitment to full personhood for marginal persons must provide for basic human needs that are consistent with one's capacity to flourish; access to adequate, affordable health care is one such need. A third point would be honoring human dignity, which means taking seriously what people on the margins *know* about their lives and what they need; similarly, it means creating social policy that honors relationality. Concretely, such policy would not, for example, create conditions wherein the personhood of the fetus or embryo is pitted against the personhood of the woman or mother; nor would it create conditions wherein one group of have-nots is pitted against another for access to the same resources or goods. A final and related point by which to assess policy reflective of full personhood for women and other marginals is one that refuses pernicious dichotomies that undergird much of our current ideology on which social policy is based.

In her work on sexual ethics, Beverly Harrison (1985) established a clear connection between the denigration and devaluation of the female body and the misogynist history of the Christian churches, particularly the Roman Catholic Church. The story is by now familiar to most of us: the dichotomies of the early Christian worldview—male-female, mind-body, spirit-nature, and so on—became naturalized and "blessed" by both church and society as the natural order of things leaving, of course, woman on the down side of the equation.

Leaping across the centuries, we know that this dichotomous worldview was instantiated in the nineteenth century in the form of the public-private split, where the cult of true womanhood means that the "good" middle-class woman confines herself to hearth and home, instilling moral virtues in her children, thus freeing the rational, middle-class man for the public realm of commerce, ideas, and politics. It further frees him from responsibility for the realm of the moral, an escape hatch he is handed even today in the making of health care policy. Indeed, women such as Susan B. Anthony were vilified for their refusal to comply with this "natural" sense of women's place.

In this third millennium, it is a cause of some concern to me that what Beverly Harrison told us in the salad days of feminist ethics is still the case—institutional social policies and practices are undergirded by a

deep mistrust of the moral agency of women and hence do not generally reflect an underlying respect for women as persons, often pitting the
personhood of the fetus or embryo against the personhood of women
themselves (Harrison 1995, 115–134). Perhaps even more pernicious,
as I alluded, such policies can also pit groups of women against each
another, often on the basis of access to technologies that are cost-
prohibitive for all but those with disposable income. "We have a long
way to go," Harrison augered, "before the sanctity of human life will
include genuine regard and concern for every female already born, and
no social policy discussion that obscures this fact deserves to be called
moral" (1985, 115). To that I would add that we have a long way to go
before the sanctity of human life will include genuine regard for the poor
and persons of color.

I suggest that this worldview with its split between public and private realms, and a correlative mind-body, spirit-nature split, constitutes
the basis for current policy positions on reproductive technologies. The
irony is that the "proper" realms for women have become inverted: it
is the *public* realm of civil society once denied women that claims to
offer us protection *against the private* and predatory realm of the free
market—no safe haven for women and the poor. But even in the public
realm, it may be observed that the personhood of women takes a back-
seat to concerns about the embryo and fetus, reflecting that naturalized
worldview of early Christianity and, in fact, adding to old dichotomies
newer ones of embryo-woman and fetus-woman. For instance, one of
the hES cell policy battles concerns the extent to which taxpayer dollars
ought to fund research on embryos and fetuses, and to what extent, therefore, women ought to be protected from engaging in such research either
as donors or consumers.

There is mistrust of women's moral agency that is perhaps easier to
see in the realm of reproductive rights and the backlash against it by the
Religious Right than it is to discern patterns of oppression in research,
development, funding, and marketing of the products of genetic technologies. Nonetheless, it is there, as this comment from Representative
Chistopher Dodd (D-Conn.) suggests. Dodd, protesting President
Clinton's revisiting the ban on federal funding for embryo research,
insisted that stem cells are "obtained from human beings ruthlessly killed

in the first weeks of life [so that] to speak of ethical safeguards in this context is a mockery when the research itself depends on the mutilation of children" (Cimons 1999). In this remark we see lack of understanding of science at work; it is not clear whether Dodd refers to stem cells derived from embryos or from cadaveric fetuses, or whether he elides the two in his reference to "mutilated children." We also see lack of regard for how any of these issues affect women and the poor, as Dodd focuses his primary concern on the premoral status of embryonic stem cells that actually do not have the potential to become full human beings (see chapters by Thomson and Okarma), even as he implies that women who have abortions have "mutilated" their "children." Dodd's comment is not taken from a diatribe against procreative choice, but against stem cell research.

What I am suggesting is that Harrison's initial insight holds true in the realm of biotechnology and can be applied to this issue of the allowable uses of embryonic stem (ES) and embryonic germ (EG) cells. The current practice of separating the ethics and practices of the private sector from those of the public sector reflects this mistrust of the moral agency of women, along with disregard for the lives of the have-nots. Rooted in the original ideology of separation between public and private spheres, government policy in this arena will inevitably be conflicted in its application, and this cannot bode well for women and those on the margins.

How does this claim stack up against proposed government policy on stem cell research and its implications for private sector profits? Let us see.

### Commodification Issues

By way of background, the NBAC was charged by President Clinton with defining the ethical uses of ES and EG cells for federally funded research, which it did in September of 1999. What are the recommended guidelines for the ethical use of ES cells by federally funded researchers? Although there are thirteen guidelines, my comments are confined only to those that bear on the question at hand—ethical uses of the policy with respect to women and other marginals.

The ethical limits of hES research are circumscribed by two conditions relevant to obtaining such cells. In other words, NBAC holds that research is permissible if conducted on one of two possible sources for hES and EG cells: embryos from fertility clinics, as opposed to those created for the purpose of research (yet another dichotomy known as the created-discarded distinction); and cadaveric as opposed to live fetuses (NBAC 1999, 68–71). Two sources ruled out by the commission—ES cells from embryos made by somatic cell nuclear transfer into oocytes, and embryos created solely for research purposes using in vitro fertilization (IVF)—interest me because both have ethical implications for women that should not be ignored simply because the commission ruled out their use for funding in the public sector. I will come back to this.

Although federal policy is evolving as I write (partly in light of recommendations from NBAC and the American Academy for the Advancement of Science [AAAS]), at the moment it still seems to be the case that federal funding is limited to research on already discarded embryos, ones that federally funded researchers did not derive, but were obtained for their use, presumably by someone in the private sector where derivation is allowable. This derivation versus use distinction (another dichotomy) has been criticized by some ethicists (see the chapter by Parens), and NBAC itself recommended that it not form the basis for federal policy on this issue. Instead, it recommends federal funding for research on embryos, both derived and used (NBAC 1999).

However, since January 1999 the National Institutes of Health (NIH) has been pursuing a policy of applying the federal ban "only to research involving the derivation of ES cells from human embryos but not to research involving only the use of ES cells" (NBAC 1999, 70). Or put in the affirmative, NIH through the Department of Health and Human Services (DHHS) permits funding for research on embryos that were not derived by NIH-funded scientists, which, as NBAC points out, might solve the legal issues, but not the ethical ones (NBAC 1999, 71). Note that the legally allowable uses of publicly funded research depends on maintaining this public-private split: as long as the publicly objectionable work is done in the private sector, the public sector can justifiably dedicate taxpayer monies to carrying out research on embryos and fetuses for the greater public good.

As Congress, federal agencies, and the corporate sector all push to clarify what constitutes allowable public-sector funding, it is worth recalling a distinction made by NBAC (1999, 69).

In the United States, moral disputes—especially those concerning certain practices in the area of human reproduction—are sometimes resolved by denying federal funding for those practices (e.g., elective abortions), while not interfering with the practice in the private sector. In this case, investigative embryo research guided only by self-regulation is a widespread practice in the private sector, and the ban on embryo research has served to discourage the development of a coherent public policy, not only regarding embryo research but also regarding health research more generally.

I would extend NBAC's observation by arguing that a policy of settling moral disputes by outlawing public funding of embryo research while fully permitting it in the private sector, is not simply an incoherent public policy on health research; it is a morally conflicted one. Such policies indicate the depth of the dichotomous worldview that gives priority to concerns of the dominant partner in any dichotomous pairing—the private sector over the public, men over women, embryos and fetuses over women, haves over have-nots.

I maintain that as long as our government sanctions separation of public and private sectors with respect to biotechnology, the proper role of the public sector with respect to research ought to be advocacy for have-nots, since the role of research in the private sector is clearly to maximize profit for investors. Regrettably, this research as advocacy position is nowhere present in the policies of this or previous administrations.

Whereas the Clinton administration is to be lauded for its use of NBAC, it does not have much to be proud of with regard to a policy of publicly funded research as advocacy for have-nots, since its record reflects instead the ideology of neoclassical economic theory. Indeed, during Clinton's presidency, for example, the number of corporate mega-mergers grew at a far greater rate than during the Reagan and the Bush administrations combined, indicating that maximized profit for the private sector is and has been, if not the end, the means of that administration's policy. We are likely to see a similar emphasis from the new Bush administration. Particularly with respect to biotechnology and the pharmaceutical industry, this has not had benign effects on consumers. An article on the pharmaceutical industry reported that the battle over corporate mega-mergers and the drive for increasing profits in drug therapies is likely to increase the problem of access for those who need but cannot afford such therapies. As a former Food and Drug Administration official commented, "They [the drugs] will increasingly treat disease. But they will also increase the disparity between the haves and the have-nots" (Rosenbaum 1999).

From a justice point of view, it is important that both ethicists and the general public scrutinize the consequences of applying federal guidelines for ethical research in science to the public sector alone. The concrete effect of legislating safeguards that apply only in the public sphere means that if the private sector wishes to pursue embryo creation by in vitro fertilization for research alone (disallowed for government funding in NBAC's recommendations), it may do so; if the private sector wishes to pursue the limits of cloning technology to create embryonic stem cells that are histocompatible (also disallowed in the public sector), it is free to do so, federal guidelines notwithstanding. If the private sector wishes to pay women whatever the market will bear for their eggs, it may do this, too; perhaps not with moral approbation, but certainly free from most legal restriction (NBAC 1999).

If market forces dictate that the demand for oocytes used in reproduction is greater than the demand for oocytes used in research, the price for the latter will be lower than that for the former. We have only to recall the advertisement taken out in several Ivy League student newspapers a couple of years ago seeking white women of high intelligence to whom parents-to-be would pay $50,000 for their donated eggs in a private fertility clinic. I am told by a colleague at an Ivy League university that a year later ads appeared in the student paper seeking specific kinds of egg donors with a $40,000 payment attached. Presumably, the market has driven down the demand for Ivy League women donors from the prior year's level! Neither must we overlook Internet auction attempts, however specious, to auction off a kidney—and now the eggs of models—to the highest bidder (Goldberg 1999; Millions Check in 1999; Online Shoppers 1999).

If some body parts are worth more than others, and in this case, if some eggs are worth more than others, as Alpers and Lo (1995) suggested, any two-tiered system of egg donation creates an inequitable and unethical situation of supply and demand. It pits two groups of women against each other in terms of market desirability on the basis of market-supported eugenics.

The eggs of well-educated Caucasians at Ivy League universities are obviously worth considerably more on the reproductive-fertility market than eggs of non-Caucasian, less educated, nonaffluent women. Might the eggs of these less valued women become a future source of research-only oocytes in the private sector? Eggs destined for laboratory research could be viewed as disposable and therefore likely to command far less than eggs used for implantation. It is already the case that the market pay-out for donors varies widely, depending on the particular character-istics sought and whatever price the market will bear. Can either situa-tion be good for women? Only the most libertarian among us would answer in the affirmative. My point is that although NBAC is concerned to keep such a situation at bay, it also recognizes that its hands are tied to the extent that we insist on this public-private distinction. Above all else, Americans are loath to regulate the market. It is one of our most enduring myths that what's good for General Motors (or Microsoft or Pfizer) is good for America.

To its credit, NBAC addressed the issue of potential commodification in its report, as well as the need fully to inform women who are oocyte donors and couples deciding to donate excess embryos to research, to guard against potential coercion by doctors or fertility clinics who might be tempted to encourage overproduction of oocytes and embryos. "Potential donors," the commissioners wrote, "should be asked to pro-vide embryos for research only if they have decided to have those em-bryos discarded instead of donating them to another couple or storing them" (NBAC 1999, 72). Beyond this, the commission recommended a six-part checklist be used by the entity seeking the donation to facilitate fully informed consent free from coercion.

A related issue that has the potential for harm to donors and others is downstream commercialization. In simplest terms it is what happens to the embryo once gamete donors have relinquished claims to it. Actually, the process originates with the oocyte donor and progresses down a con-tinuum until the embryo reaches applications with the potential for tremendous profit. "The subject of commercialization is a potentially important one, affecting both researchers who must acquire embryos from for-profit IVF clinics or other sources and downstream users who may develop derivative, commercial applications from basic embryolog-ical and stem cell research" (NBAC 1999, 36). Issues include whether the donors must be "informed about the nature of and potential commercial uses of the biological materials they donate," and NBAC answered in the affirmative: in the public sector donors must be so informed. The problem is that in the private sector such regulations are not binding, but merely voluntary and despite the fact that although "... state statutes on organ transplantation now typically prohibit sale of human organs or parts ... none includes language likely to impede research involving human embryos" (NBAC 1999, 47). Therefore, downstream commercialization is a potent and problematic issue. How to safeguard it ethically and how to keep women from potential exploitation is the rub. The potential profitability of cell lines derived from donated embryos is huge given the promise of regenerative medicine. Indeed, it is why Geron Corporation invested so heavily in hES cell research.

Clearly, NBAC calls for prohibition of commerce in embryos and fetal tissue, as its recommendation 7 states: "Embryos and cadaveric fetal tissue should not be bought or sold." However, this recommendation has teeth only in the public sector, and here it would be possible to force compliance of corporate interests, but only if these interests intend to make their cell lines available to publicly funded researchers. Then, the corporation "must submit its derivation protocol(s) to the same oversight and review process recommended for the public sector ..." (NBAC 1999, 108). Realizing that its recommendations have no enforceability beyond the reach of the federal purse, NBAC nonetheless added to its report *Recommendation 11: Voluntary Actions by Private Sponsors of Research that Would Be Eligible for Federal Funding*, wherein the private sector is "encouraged to adopt voluntarily the applicable recommendations of this report" (NBAC 1999, 108).

Here NBAC is essentially forced to admit that all of its best efforts at navigating the murky ethical waters of hES cell research can, in fact, be undercut by this dichotomy between public and private sectors where two sets of standards apply: a complex and regulatory one for public funding, and no regulations at all for the private sector. In my view, this reinforces what I previously asserted is the chief moral value of American culture under late advanced capitalism: thou shalt not regulate the free market. In any case, NBAC virtually pleads for voluntary compliance with its ethical standards and safeguards, even as it is forced to recognize that "Some of the recommendations made in this context—such as the requirement for separating the decision by a woman to cease such treatment when embryos still remain and her decision to donate those embryos to research—simply do not apply to efforts to derive ES cells from embryos created (whether by IVF or somatic cell nuclear transfer) solely for research purposes, activities that might be pursued in the private sector" (NBAC 1999, 109). Its solution is to encourage professional trade associations and societies to develop ethical standards that might have the effect of bringing its private sector members into compliance with the ethical safeguards of its report.

## Conclusion

What is so compelling about moving forward with research on hES cells is the promise that it contains the potential for therapies that "will serve to relieve human suffering," as NBAC put it. But a feminist ethical analysis has to ask, whose suffering? and at whose expense?

It is true that women, the poor, persons of color, and marginals could benefit from the regenerative medical therapies and drug therapies heralded by hES cell and EG cell research, but it is not at all likely that they will be the ones who do benefit. Such therapies, when they are perfected, are likely to be cost-prohibitive for all but the wealthy and the well-insured, assuming that insurance companies agree to such coverage, a big assumption in any case. The poor, who are largely female, and most persons of color will simply be marginalized from these therapies, even as it is possible that their eggs are commercialized downstream for profit.

Finally, even if access to regenerative medicine were equitably shared, the question is never addressed of whether we *ought* to expend precious resources in this arena while daily, the numbers of persons without access to basic health care grows. How will hES cell research and the public monies devoted to it compare with funding and research dedicated to issues that matter for women and persons of color? I think we will have to answer these questions, and the concerns of African-American women, who tell us what health care priorities mean to them. In the words of the Woman of Color Health Partnership (1994):

We are still an embattled people beset with life-and-death issues. Black America is under siege. Drugs, the scourge of our community, are wiping out one, two, three generations. We are killing ourselves and each other. Rape and other unspeakable acts of violence are becoming sickeningly commonplace. Babies linger on death's door, at risk at birth: born addicted to crack and cocaine; born underweight and undernourished; born AIDS-infected. An ever-growing number of our children are being abandoned, being mentally, physically, spiritually abused. Homelessness, hunger, unemployment run rife. Poverty grows. Our people cry out in desperation, anger, and need.

When people on the margins are beset with life-and-death issues, ES cell research and other genetic technologies can appear to be luxury items with little chance of reaching these individuals. More, the risk is that public funding, already scarce for the real health concerns of African-Americans, for example, will be diverted away from those life-and-death needs. It is already the case that persons of color suffer most from the current piecemeal health care financing and insurance system. In 1998, the latest year for which such figures are available, of 43 million Americans without health insurance coverage, fully 58% were black or Hispanic (Toner 1999).

Even if public funding focused attention on the health needs of persons on the margin, the private sector has neither obligation nor incentive to do so. There simply is not the chance for return on investment that can be reaped through genetic technologies and pharmaceuticals targeted to those with the ability to pay. This is the legacy of the "free market."

I will close this chapter as I began it, with reference to Susan Sherwin's feminist imperative for research—that it be evaluated "not only in terms of its effects on the subjects of the experiment [the moral status of the embryo] but also in terms of its connection with existing patterns of oppression and domination in society." Although I am not advocating an end to hES cell and EG cell research—the potential applications hold great promise for humankind—when analyzed from the light of patterns of domination and oppression it does not fare well for the have-nots. My fear is that it will be yet another step toward increasing marginalization of the many on behalf of the few.

My analysis will doubtless be unpopular with those who see in science the great march toward endless human progress, and who see in that progress the potential for lucrative profits. I contend that what we need in this debate, as in so many other questions of science and public health dollars, is evidence of a commitment to the have-nots. In our time, such a stance requires genuine moral courage on the part of policy makers, for there is little to be gained in terms of things measured by most politicians. A coherent public policy will not give into the ethical split that results from a dichotomous worldview wherein the lives of those on the margins are subject, on one hand, to moral whims of legislators and on the other hand to amoral whims of the private sector economy.

No doubt we need a full-blown analysis of distributive justice on this issue. My task, however, has been more modest. It has been to suggest what a feminist ethical analysis of the hES cell debate might entail. At the moment, until such time as placement of our dollars, both public and private, reflects more equitable distribution of benefits and burdens, a feminist ethical analysis reminds us to pause. In so doing, we would do well to subject public policy in question to the scrutiny of a moral litmus test that ensures the least well-off among us that they will be as likely to benefit as the most advantaged. Justice demands no less.

## Notes

1. Although my concern is for justice, I do not intend to mount a thorough critique of distributive justice; only to raise questions about it from a feminist perspective. Such an analysis would be welcome, but is beyond the scope of this chapter.

2. See, for example, hostile takeover bids in the pharmaceutical industry among Warner-Lambert, Pfizer, and American Home Products. See also the Federal Trade Commission's investigation into the "vitamin cartel" of many of these same transnationals and their price-fixing of vitamins. See also repeal of the Glass-Steagall Act and the Bank Holding Company Act, which allows bank, securities firms, and insurance companies to merge (Reuters 1999).

## References

Alpers, A. and Lo, B. 1995. Commodification and commercialization in human embryo research. *Stanford Law and Policy Review* 6: 2.

Cimons, M. 1999. NIH says it will fund human stem cell studies. *Los Angeles Times*, January 20, Part A section.

Goldberg, C. 1999. Selling fashion models' eggs raises ethics issues. *New York Times*, October 23. Available at http://www.nytimes.com/library/tech/99/10/biztech/articles/23eggs.html.

Harrison, B. 1985. *Making the Connections: Essays in Feminist Social Ethics.* (Carol Robb, ed.). Boston: Beacon Press.

Millions check in on internet auction of models' ovarian eggs. 1999. Available at wysiwyg://32/http://cnn.com/HEALTH/women/9910/25/models.eggs.ap/.

National Bioethics Advisory Commission. 1999. *Ethical Issues in Human Stem Cell Research*. Vol. I. *Report and Recommendations of the National Bioethics Advisory Commission*. Rockville, MD: National Bioethics Advisory Commission.

Online shoppers bid millions for human kidney. 1999. Available at http://www.cnn.com/TECH/computing/9909/03/ebay.kidney/.

Reuters. 1999. Clinton signs legislation overhauling banking laws. *New York Times*, November 13, p. B3.

Rosenbaum, D. 1999. The gathering storm over prescription drugs. *New York Times Week in Review*, Section 4, p. 1.

Sherwin, S. 1992. *No Longer Patient: Feminist Ethics and Health Care*. Philadelphia: Temple University Press.

Toner, R. 1999. Rx redux: Fevered issue. Second opinion. *New York Times Week in Review*, Section 4, p. 1.

Women of Color Partnership. 1994. We remember: African-Americans are for reproductive freedom. Washington, DC: Religious Coalition for Reproductive Choice.

Young, I. M. 1990. *Justice and the Politics of Difference*. Princeton, NJ: Princeton University Press.

# III

## Angles of Vision

# 8

## Stem Cell Research—A Jewish Perspective[1]

Elliot N. Dorff

**Fundamental Theological Assumptions[2]**

I begin with some basic theological assumptions. First, the Jewish tradition uses both theology and law to discern what God wants of us. No legal theory that ignores the theological convictions of Judaism is adequate to the task, for such theories lead to blind legalism without a sense of the law's context or purpose. Conversely, no theology that ignores Jewish law can speak authoritatively for the Jewish tradition, for Judaism places great trust in law as a means to discriminate moral differences in similar cases, thus giving us moral guidance. My understanding of Judaism's perspective on stem cell research will, and must, draw on both sources.

Our bodies belong to God; we have them on loan during our life. God, as owner, can and does impose conditions on our use of our bodies. Among those is the requirement that we seek to preserve our life and health.

The Jewish tradition accepts both natural and artificial means to overcome illness. Physicians are agents and partners of God in the act of healing. Thus the fact that human beings created a specific therapy rather than finding it in nature does not impugn its legitimacy. On the contrary, we have a duty to God to develop and use any therapies that can aid us in taking care of our bodies. At the same time, all human beings, regardless of their levels of ability and disability, are created in the image of God and are to be valued as such.

We are not God. We are not omniscient, as God is, and so we must take whatever precautions we can to ensure that our actions do not harm

ourselves or our world in the very effort to improve them. A certain epis-
temological humility, in other words, must pervade whatever we do,
especially when we are pushing the scientific envelope, as we are in stem
cell research. We are, as Genesis says (2:15), supposed to work the world
*and* preserve it; it is that balance that is our divine duty.

### Jewish Views of Genetic Materials

Since doing research on human embryonic germ cells may involve
procuring them from aborted fetuses, the status of abortion within
Judaism immediately arises. By and large, abortion is forbidden. During
most of its gestational development, the fetus is seen as "the thigh of its
mother," and neither men nor women may amputate their thigh at will
because that would be injuring their bodies. On the other hand, if the
thigh turns gangrenous, both men and women have the positive duty to
have it amputated to save their lives. Similarly, if the woman's life or
health is at stake, an abortion *must* be performed to save her life or her
physical or mental health, for she is without question a full-fledged
human being with all the protections of Jewish law, whereas the fetus is
still only part of her body. When a risk to the woman is elevated beyond
that of normal pregnancy but not so much as to constitute a clear threat
to her life or health, abortion is permitted but not required; that is an
assessment that the woman should make in consultation with her physician. Some authorities would also permit abortions when genetic testing
indicates that the fetus will suffer from a terminal condition such as Tay-
Sachs disease or from serious malformations.[3]

   The upshot of the Jewish stance on abortion, then, is that *if* a fetus
was aborted for legitimate reasons under Jewish law, it may be used to
advance our efforts to preserve the life and health of others. In general,
when a person dies, we must show honor to God's body by burying it as
soon after death as possible. To benefit the lives of others, though,
autopsies may be performed when the cause of death is not fully understood, and organ transplants are allowed to enable other people to live
(Dorff 1998). The fetus, as I have said, does not have the status of a full-
fledged human being. Therefore, if we can use the bodies of human
beings to enable others to live, how much the more so may we use a part
of a body—in this case the fetus—for that purpose. This all presumes,

however, that the fetus was aborted for good and sufficient reason within the parameters of Jewish law.

Stem cells for research purposes can be procured not only from aborted fetuses, but also from donated sperm and eggs mixed together and cultured in a petri dish. Genetic materials outside the uterus have no legal status in Jewish law, for they are not part of a human being until implanted in a woman's womb; even then, during the first forty days of gestation, their status is, according to the Babylonian Talmud, "as if they were simply water."[4] Abortion is still prohibited during that time except for therapeutic purposes, for in the uterus such gametes have the potential of growing into a human being, but outside the womb, at least as of now, they have no such potential. As a result, frozen embryos may be discarded or used for reasonable purposes, and so stem cells may be procured from them.

**Other Factors in this Decision**

Given that the materials for stem cell research can be procured in permissible ways, the technology itself is morally neutral. It gains its moral valence on the basis of what we do with it.

The question, then, reduces to a risk-benefit analysis of stem cell research. Articles in a *Hastings Center Report* (March-April, 1999, 30–48) raise questions to be considered in such an analysis, and I will not discuss them here. I want to note only two things about them from a Jewish perspective:

First, the Jewish tradition sees the provision of health care as a communal responsibility, so the justice arguments discussed there have a special resonance for me. Especially since much of the basic science in this area was funded by the government, the government has the right to require private companies to provide applications of that science to those who cannot afford them at reduced rates or, if necessary, for free. At the same time, our tradition does not demand socialism, and for many good reasons we in the United States have adopted a modified capitalistic system of economics. The trick is to balance access to applications of the new technology with the legitimate right of a private company to make a profit on its efforts to develop and market applications of stem cell research.

Second, the potential of stem cell research for creating organs for transplantation and cures for diseases is, at least in theory, both awesome and hopeful. Indeed, in light of our divine mandate to seek to maintain life and health, one might even contend that from a Jewish perspective we have a *duty* to proceed with that research. As difficult as it may be, we must draw a clear line between uses of this or any other technology for cure, which are to be applauded, as against uses for enhancement, which must be approached with extreme caution. Jews have been the brunt of campaigns of positive eugenics both here in the United States and in Nazi Germany (Gould 1996; Annas et al. 1992), so we are especially sensitive to the dangers in creating a model human being that is to be replicated through the genetic engineering that stem cell applications will involve. Moreover, when Jews see a disabled human being, we are not to recoil from the disability or count our blessings for not being afflicted in that way; we are rather commanded to recite a blessing thanking God for making people different.[5] In light of the view that all human beings are created in the image of God, regardless of their levels of ability or disability, it is imperative from a Jewish perspective that applications of stem cell research be used for cure and not for enhancement.

### Recommendation

My recommendation is that we take the steps necessary to advance stem cell research and its applications in an effort to take advantage of its great potential for good. We should do so, however, with restrictions to enable access to its applications to all Americans who need it and to prohibit applications intended to make all human beings into any particular model of human excellence. We should instead seek to cure diseases through this technology and to appreciate the variety of God's creatures.

### Notes

1. This essay is adapted from Rabbi Dorff's testimony before NBAC May 7, 1999.

2. For more on these and other fundamental assumptions of Jewish medical ethics, and for the Jewish sources that express these convictions, see Dorff (1998).

3. For more on the Jewish stance on abortion, together with biblical and rabbinic sources that state that stance, see Dorff (1998) and Feldman (1968, 1973).

4. Rabbi Immanuel Jakobovits notes that forty days in talmudic terms may mean just under two months in our modern way of calculating gestation, since rabbis counted from the time of the first missed menstrual flow whereas we count from the time of conception, approximately two weeks earlier (Jakobovits 1959, 1975, 275).

5. For a thorough discussion of this blessing and concept in Jewish tradition, see Astor (1985).

# References

Annas, G. J. and Grodin, M. A. 1992. *The Nazi Doctors and the Nuremberg Code: Human Rights in Human Experimentation*. New York: Oxford University Press.

Astor, C. 1985. *Who Makes People Different: Jewish Perspectives on the Disabled*. New York: United Synagogue of America.

Dorff, E. N. 1998. *Matters of Life and Death: A Jewish Approach to Modern Medical Ethics* Philadelphia: Jewish Publication Society.

Feldman, D. M. 1968. *Birth Control in Jewish Law*. New York: New York University Press. Reprinted in 1973 as *Marital Relations, Abortion, and Birth Control in Jewish Law*. New York: Schocken.

Gould, S. J. 1996 *The Mismeasure of Man*. New York. Norton.

*Hastings Center Report*. March-April, 1999.

Jakobovits, I. 1959, 1975. *Jewish Medical Ethics*. New York: Bloch.

# 9

# The Ethics of the Eighth Day: Jewish Bioethics and Research on Human Embryonic Stem Cells

Laurie Zoloth

For scholars in the Jewish tradition, new science presents new challenges. We are increasingly and urgently asked about the permissibility, the telos, the moral meaning, and the appropriate limits of remarkable advances in biotechnology and genetic medical interventions that are fundamentally changing our most basic understanding of what it means to be human, of what the proper limits ought to be on research, and on the moral status of the essential component parts of human biology itself. This chapter is intended as an account of the ethical analysis of Jewish bioethics in its first response to the emerging research about human embryonic stem cells. It is both preliminary and partial in that it is an except from longer work, and that issues in genetics are only recently the focus of Jewish ethics (Zoloth 2000). Furthermore, for the Jewish ethical-legal tradition (*halachah*), which functions methodologically as a discursive community in which justification is created by the force of moral suasion, no single authoritative voice or one particular council of authority speaks for the entire tradition or the community. Jewish reasoning is a series of open-ended arguments intended to include the broad and creative use of history, text, and culture, with many interrupting voices representing competing narratives. Hence, in confronting emerging ethical issues, what will serve best in framing a coherent Jewish response is the widest possible call for inquiry to delineate the types of questions that further work will have to address.

At stake in this reasoning is finding cases that, although not identical to ours, have distinguishing moral appeals that might be similar. In this, *halachic* reasoning is a form of linguistic, definitional analysis in which parties to the debate seek epistemologic commonality as a first step to social policy. Furthermore, for Jewish ethics, the framing questions are those of obligations, duties, and just relationships to the other, rather than protection of rights, privacy, or ownership of the autonomous self. Since much of our thinking in contemporary American bioethics is rights based and relies on a model of intricate semilegal contracts carefully made between autonomous and anonymous strangers, the idea of centering our obligations rather than worrying about our rights can seem simple-minded or naive. But the other-regarding, binding gesture, this commanded act of justice, responsibility itself, is the first premise of Jewish ethics (Levinas 1990).

Jewish consideration of issues in bioethics is, of course, textually based, and based in the casuistry of halachah in which specific considerations are addressed by textual recourse.[1] Halachic reflection on all innovative scientific research is constrained by the fact that none of the specific issues raised by new technology is directly addressed by Talmudic conversations compiled in the first centuries of the common era, or in the elaborate medieval commentary that carries the most considerable weight in the classic tradition. An important caveat is that the new terrain on which we now find ourselves bears scant analogy to the terrain of the rabbinic world. The biology of the Talmud was still couched in terms later altered or reframed, gamete reproduction was still not fully understood, and microbiologic techniques were not even imagined. Moreover, in researching halachic conversations that touch on this arena, we can note that what the rabbinic culture understood as central is not necessarily what moderns consider most salient. For example, whereas moderns are worried lest we "play God," the rabbis were concerned that we act *more* like God might in many ethical and social-political arenas, as in helping the poor, creating justice, and healing the sick. But Jewish ethics does not proceed without questions, so let me raise them.

### In General, There Are Three Categories of Debate

Research on stem cells, on the possibility of manipulating them, pushing them toward differentiation, or from pluripotency to totipotency, or growing vast amounts of them all raise issues of definition and meaning. Are human persons collections of potentially deconstructable and dismantleable other parts, or even other selves?[2] What is the moral status or the ontologic nature of the self; the intent and scope of medical intervention; what constitutes disease and what normalcy, and the relationship between God and human partners?

The next question, important in a religious legal system such as Judaism, is whether technical aspects of the complex manipulation required are themselves permitted. Informed consent, use of advanced reproductive technology (ART) and its attendant contracts, and limits on applications and participants all have to be addressed.

The last set of questions, and one that is critically important in Jewish thought, concerns issues of justice, access, distribution, and implications of the work on the human community in which we share an altered medical and social universe.

### The Problems of Telos

The first Jewish responses to human embryonic stem (hES) cell and embryonic germ (EG) cell research seem to indicate general sanguinity with the procedure, framed as breakthrough medical therapy for life-threatening conditions. This general response is based on the clear mandate in Jewish texts to save life whenever possible, even if it requires violating or suspending other commanded acts (*mitzvot*). To save even one life, the *halachah* states, it is permissible, and in fact it is mandated, that all other *mitzvot* can be abrogated (except for prohibitions against murder, adultery, and idolatry). This category of response stems largely from the defining moment in the Talmud in which rabbinic authorities debate whether one can violate the mandate to rest and sanctify the Sabbath to rescue a man trapped in the rubble of a collapsed building. From this vivid (and graphically obvious) source text springs a whole set of cases that are defined as like being trapped—by illness, catastrophe, hunger, war, or threat. This has provided the warrant text for virtually all experimental therapy, including genetic research. Hence, even otherwise proscribed actions (e.g., taking organs of the dead) are permitted if a life can be reliably saved. Jewish medical ethics is nearly entirely constructed around this principle of *pikuach nefesh* (to save a life) (Dorff 1999).

This consideration can be proposed about nearly all the technologies that are suggested by this research. If full use were possible for this tissue, millions of persons would be afforded years of productive life, and repairing, patching, transfusing, and replacing damaged tissue would alleviate human suffering without altering the essential self of the recipient.

### The Problems of Origins and Moral Status: What Age Is the Embryo?

But all of this reasoning changes if we consider the embryo (which is destroyed to get to the inner cellular mass) to be a person with full moral status. By moral status, we mean how we describe the standing of an entity relative to other moral agents, and the obligations and relationships that other moral agents have toward this entity[3] (Warren 1998). If hES cells are understood as only tissue or as organic nonhuman life forms, it might be permissible to use them even instrumentally for very compelling reasons and just ends. If embryos or cells have full human moral standing, our obligation toward them shifts sharply.[4] Many commentators in Roman Catholic and Protestant theological traditions indeed see the embryo as a fully human person (Pellegrino 2000).

Whereas moral status of embryonic tissue is the threshold question for many religious traditions, the Jewish position is that this is of secondary importance to the debate, to be noted after the life-saving consequences of this technology are established. Like nearly all discourse in this field, much Jewish understanding of moral status derives from the abortion debate, in which the embryo and fetus have a developmental status relative to their gestational age. At stake is whether the fetus is an independent entity or part of the body of the mother (*ubar yerickh imo*). The biblical text (Exodus 21:22) that grounds the literature is as follows:

If two men fight, and wound a women who is pregnant (and is standing nearby) so that her fruit be expelled, but no harm shall befall (her) then he shall be fined as her husband assess, and the matter placed before the judges. But if harm befall her, then shalt give life for life.

The text is understood to mean that if the women herself is not harmed, the only harm, loss of the pregnancy, is of lesser importance and can be made whole by monetary compensation, unlike the taking of a human life.

Central to the understanding of embryology in the Talmud and subsequent halachic responsa is that before the fortieth day after conception, the developing fetus is to be considered "like water." Rabbis were close observers of fetal development because it fell within their purvey to examine all genital emissions, to answer questions of *niddah* (the period during the monthly cycle that a husband and wife are not permitted sexual relations), and use of *mikvah* (ritual immersion after *niddah*). At stake here is the understanding that the relationship of a woman to her community was closely tied to the moral status of the delivered fetus: was this a stillbirth or a late menstrual period? Would the women be in *niddah* for 14 days or 6 weeks?

In that capacity, discussions surround the nature and character of the contents of the womb at various stages of development. Other considerations, such as quickening, and external visual changes in a woman's body also warrant different social responses and different consideration of the pregnancy. This developmental understanding of moral status is not limited to the fetus. There is ample precedence for the rabbinic understanding of changing obligations, even life-saving obligations, based on the temporal standing of the human person. Unlike other traditions, liminal times and moral status questions exist not only at the beginning but also at the end of life, and well-established norms permit instrumental consideration of an entity, clearly a human person, and clearly alive, based solely on this understanding of developmental moral statuses (Feldman 1995).

Let us turn to classic examples. When a person is in a state called *terefah* (having a fatal organic disease) our obligations to save his life and his life relative to others is altered. Texts discuss categories of persons to whom one might be differently obligated to protect in a crisis such as a hostage-taking[6] (Dorff 1999). Since the death of a *terefah* is by definition inevitable, killing this person does not count as murder in quite the same way, nor is the civil obligation toward him exactly the same as if he were not fatally ill. This critical liminal state involves a highly nuanced view of personhood.[7] Rabbis struggled to define these states and used different vocabularies in an attempt to describe with accuracy a difficult and essentially mysterious boundary of human life.

A parallel in rabbinic categorization is the beginning of personhood and the debate that surrounds abortion. Jewish law suggests a liminal status for the fetus and, exempts from the death penalty, its destruction. Subsequent rabbinic commentary regards the fetus as "a part of the women's body" until the moment at which the head or the greater part of the breech is delivered out of the birth canal. Up until that moment, a pregnancy can be terminated and the fetus allowed to die to save the life or health (mental or physical) of the woman. After infants are born, their moral status is still in a process of development, albeit of a less dramatic nature. Children are not named or admitted to community (public) membership until the eighth day of life, and if a child dies before the thirtieth day of life, the necessary rituals of death are not preformed (*shiva* is not observed and the *kaddish* is not said for the requisite year of mourning). All of these considerations frame a Jewish view of the moral status of the preimplantation embryo: a non-ensouled entity that is deserving of special consideration and respect, but not a human person within the mutually binding halachic system (Feldman 1968). Furthermore, if we can determine a distinction in the moral status of the embryo before and after 40 days, surely we can determine a distinction between the preimplantation and postimplantation embryo. In reflecting on possible abuses of a blastocyst that might be created by the nuclear transplant (cloning) and subsequent use of stem cell techniques, theologian and legal scholar Ze'ev Falk raised the issue of whether one could even consider the entity that is an artificially created blastocyte to be a human, since it was not created by sexual intercourse (personal communication 1998). In his reflections on this topic, Falk noted that the tissue's origin would make it distinctive from a naturally occurring pregnancy in a womb, because it is fabricated externally, out of its normal course of development. Further issues emerge about the legal status of tissue cell lines themselves: are they to be regarded as part of a women's body for as long as they exist? How does ownership accrue to them? Since rabbis did not have a halachic category for cells that can live ceaselessly and are, perhaps, capable of asexual reproduction, it will require further research to claim anything about the halachic status of this tissue.

Is the pursuit of genetic research *mandated* healing? The task of healing in Judaism is not only permitted, it is mandated; if stem cells can save a life, then not only can they be used, they must be used. This is supported and directed not only in early biblical passages ("you shall not stand idly by the blood of your neighbor," and "you shall surely return what is lost to [your neighbor]," etc.), but in numerous rabbinic texts (Dorff 1999). The general thrust of Jewish response to medical advance has been positive, even optimistic, linked to the notion that advanced scientific inquiry is a part of *tikkun olam*, the mandate to be an active partner in the world's repair and perfection. Judaism is not, after all, a nature-based religion; the very assertion of circumcision rests on the notion that the body is not sacred or immutable. No part of the body is sacred or untouchable. But disfigurement of the body (e.g., piercing, tattoos) is not permitted, and the belief that the personal body is a property that belongs to the self alone is a late and nontraditional response to medical decision making. Characteristically, "Judaism does not interfere with physician's medical prerogative, providing his considerations are purely medical in character"[8] (Jacobowitz 1959).

This permission and obligation to heal come directly from the Torah text of Exodus and Deuteronomy, as interpreted by the Talmud:

"The school of R. Ishmael taught and healed, he shall heal (Exodus 21:19). This is [the source] whence it can be derived that the authorization was granted [by God] to the physician to heal."

And further: "How do we know [that one must save his neighbor from] the loss of himself? From the verse: And thou shall restore him to himself" (Deut. 22:2).

A positive attitude toward medicine stresses that the recourse to prayer and faith alone is incomplete without the complete resourcefulness of which humans are capable. This capability is a God-given gift, part of the work of stewardship to which persons are entasked in Genesis.

As many commentators have noted, another text directs the general attitude of Jewish theologians toward the medical endeavor. The physician's work is legitimate and in fact obligatory, as can be seen in the following story. Rabbi Akiba and Rabbi Ishmael are walking in Jerusalem and encounter an ill person who asks them to cure him. They do, but the man is puzzled: after all, are not the rabbis transgressing the will of God, who made him sick in the first place, by curing him? They answer by asking him about his work. He is a farmer, who works in the vineyard created by God: does he not alter the world that God created by his work?

He answers them, "If I did not plow, sow, fertilize, and weed, nothing would sprout." Rabbi Akiva and Rabbi Ishmael said to him, "Foolish man.... Just as if one does not weed, fertilize, and plow, the trees will not produce fruit ... so with regard to the body. Drugs and medicaments are the fertilizer, and the physician is the tiller of soil." Thus humans have a mandate to be partners with God in creation, to repair a unfinished world (called the task of *tikkun olam*.) Dorff (1999), for example, generally acclaimed genetic engineering as "one of the wonders of modern medicine." Whereas he noted the potential for eugenic uses, "the potential benefits to our life and our health are enormous," and hence research ought to continue.

No specific texts address the use of research science specifically, although the Talmud is replete with stories about the general ability of rabbis to examine closely the abortus itself, or to observe specific medical conditions. On the other hand, no halachic texts forbid basic research either. David Bliech (1981) noted that these phenomena are characteristic of several modern problems in medicine, ones for which there are no clear textual referents. More recently, Bliech (1999) used texts that refer to the necessity to build fortification around cities. The community must build walls in the face of danger, but its obligation to protect itself against imminent danger does not extend to danger that exists in the not-yet-existing future. Thus, by extrapolation, genetics work that promises the very real chance of saving a life is an obligation to pursue even in the face of other theoretical dangers. In Bliech's view, the premise is clear: God has left humans a broken, unfinished world, and our task is to complete it by our actions.

Given such optimistic halachic responses, the nearly universal communal response to all genetic advances that can promote health and increase fertility has been enthusiastically positive in the Jewish world. The absolute mandate to heal, and firm rejection of some Christian claims that to heal is to counter God's will, is a consistent feature of normative Judaism. Furthermore, it is mandated to use the best methods available as soon as they are proved efficacious and not dangerous to the patient. Paradoxically, it might violate rabbinic premises to *stop* research if such research is life saving.

### Problems of Process

Issues of moral status dominate the discourse for moderns, but for classic rabbinic commentators, medical practices raised significant concerns as well. Attentive to needs of women over concerns for embryos, Jewish ethics asks: how is it that one obtains material for research? For example, can we use drugs to stimulate ovulation? The problem of biblical infertility is resolved on the spiritual level, but there is no prohibition against the use of all medical intervention that can help a couple achieve the commandment to raise a family, and agents to stimulate fertility are spoken of approvingly in biblical literature.[9]

But there is rabbinic concern about process: is it adultery if the sperm of another man is used inside a women's body (as in artificial insemination)? For this reason, some orthodox rabbinic sources prohibit use of donor sperm for artificial insemination. Special considerations exist even for use of the husband's sperm or, in some cases, a mixture of sperm sources to meet halachic requirements. Sperm is not to be wasted (the sin of Onan), so elaborate collection devices have been created to allow for coital stimulation and collection of sperm (Jerusalem Report 1998). Does the process of egg harvesting shame the woman? The dignity, reputation, and integrity of her body were all significant considerations for rabbinic authorities, who were deeply concerned about protecting her body from any event that would force her into shame. In this discussion, the consideration is close to the feminist stance that understands gametes as a part of a woman's self and not as property to be sold. Clearly we must reflect carefully on the informed consent process. Later texts are clear that the embryo and fetus are not the property of the husband. As such, since the fetus is considered part of the woman's body, the women's mental status must be considered carefully, as well as circumstances surrounding collection of the egg.

Even the informed consent required to donate raises ethical questions. Here we encounter the problem of a coerced and therefore unenforceable contract, a specific entity in Jewish law. It has been noted for centuries of legal debate that some contracts are not valid because they require an unnatural (in the rabbinic sense) act of imagination or will and cannot be enforced justly. A contract that is clearly not in the best interest of the person who makes it is not valid. Is the agreement to donate blastocysts or fetuses such a contract? Does consent for the use of fetal tissue or of embryos involve such contracts? Linked to this are other possibilities for source texts: perhaps in laws regarding the use of slaves or limits on the use of female slaves or of captives, one might find ideas for how we think of relationships between persons of different power involved in instrumental relationships. Here again, careful work will have to be done to determine such relational issues.

Origins of tissue raise other concerns, even if donor eggs are not used. If the cells originate from gamete (EG) cells of aborted fetuses, *halachic* considerations turn us in another direction: is it disrespectful of the dead to collect them? We must determine when the abortion is actually preformed. Timing is essential for both researcher and halachacist, since cells must be collected before they differentiate. and since rabbis understand moral status to be developmentally acquired. If collection is after forty days (of what rabbis would consider) conception, we have a new problem concerning medical use of body parts of the aborted fetus. To address this problem, I turned to the protracted debate about autopsy in the halachic literature. It seems clear that cutting, dissecting, and using fetal tissue border on prohibitions about desecration of the dead. But several factors mitigate this problem. The fetus is not the same as the stillborn child. Next, in the case of the permitted autopsy, the procedure is permitted as described above, *pikuach nefesh*. Whereas some assurances must be given that a specific life will be saved by the medical information derived from the procedure, many allow autopsies to enhance understanding of a disease process that affects a category of ill persons.

Moral status for hEG cells could be less troubling, since the cells are taken from the gamete ridge of an already dead fetus and the justification is akin to uses of other sorts of human cadaver tissues, such as skin for grafting in burn victims and kidneys for transplantation. This autopsy model yields important results in our moral theory as well. We may have qualms about the origins of the aborted fetus, and we may not like or may even abhor the circumstances of the death of the fetus, but we understand that we may use the tissue for important and good ends. In thinking about this, we may make an extreme comparison imagining the aborted fetus in exactly the same way we might allow the use of the kidneys or skin of a victim of a drive-by shooting. Use of tissue is in no way seen in the second case as an endorsement of drive-by shootings, and use of tissue in the first case is not an endorsement for abortion. Other issues of process and other persons require reflection. I will list them very briefly.

The issue of safety is critical in Jewish law. The hES and EG cells are by nature unstable. We are only now learning about their specific properties, but clearly, some of what makes them interesting could make them dangerous in ways that may not be expressed for generations. For example, the highly telemerese expressive quality of these cells means that they can proliferate and are immortal. But this is a quality shared with cancer cells. Will these cells retain this characteristic in higher percentages when used in vivo? Another question arises relative to the cells' mutability. Will implanted hES or EG cells have a high rate of mutability? How will we be able to test for such effects?

Jewish law is also concerned about the other moral actors in research. How would performing the act of harvesting aborted fetal cells and all that this entails affect the scientists involved? What must be considered to protect researchers from becoming indifferent or coarsened to the human tissue involved? How can scientists, by design removed from patients to protect the informed consent process, still act as if they are healers, motivated in ways that must inform and direct the research? How will the significant monetary incentive affect this commitment? What is the effect on society if we create a bank of canonical cell lines, considering the potential of each cell and its special status?

### Issues of Context: Is This a Just Use of Technology?

Many of Jewish law and codes are concerned with justice in an unjust world. How to create a world of just order is a clear preoccupation of the biblical and rabbinic argument about the meaning and goals of a society that lives in a covenental relationship with God. For justice to have real meaning, the civilization that is constructed will have to account for the primacy of this relationship. Can the interests of the vulnerable be heard in our debate?

In Jewish thought, the poor are to be protected not only out of a vague sense of compassion, but as a part of how the natural and agricultural world is structured. Our texts remind us that the harvest is understood to include the provision of parts of the field and parts of the yield for the poor. In fact, essential economic decisions (such as how to plant, what to harvest, and when to refrain from planting) are mediated by this consideration. Limits are placed on the entire society to ensure that the widow, orphan, and stranger are provided for with full dignity. Hence concern for the sabbatical year in which all production is suspended to allow for the use of the field by the disadvantaged, for the harvest to be organized to allow for gleaning, for the corners of the field to be proscribed for one's own use and to be reserved for the use of the poor. Technologic advances, even clever and expedient ones, cannot be permitted if persons or even animals might be unjustly used; hence concern is raised when yoking unlike animals for plowing.

Is this a good instance of *tikkun olam* or overreaching of human power? Does intuitive uneasiness at new technology arise from a sense that we wield too much knowledge that we cannot morally control? Against the normative optimism for medical research, two important texts recall a broad general caution for all technology. The first is the creation of the golem, an all-powerful humanoid creature, by manipulation of text and spells (Babylonian Talmud, Sanhedrin 65b). This theme recurs frequently in the tradition and I have noted its centrality in other works (Zoloth-Dorfman 1998). Yet as appealing as this image is to a persecuted people, we are warned of the essential error in the pursuit of this particular type of creationist research that leads to the excesses of spiritless power, unguided by faith, and ultimately dangerous.

The second text about technology is the *midrash* on the construction of the Tower of Babel. Here rabbis struggle with why construction of a joint human project is problematic, even when the ostensible reason is to "reach up to God." Finding nothing in the direct text, they describe a theoretical scene: "When a worker was killed, no one wept, but when a brick fell, all wept" (*Midrash* Rabbah).

What is occurring here? The rabbinic caution was that using humans instrumentally in a technologically impressive human project led to dismantling of the distinction between persons and things. It was this decentering of the human and reification of the thing that was the catastrophe that felled the enterprise, perhaps, suggests this text, as much as the hubris of trying to pierce heaven. It is not just that they breached a limit between what is appropriate to create and what is not, the process of the creation must be carefully mediated, with deep respect for persons over the temptations of the enterprise. Such a text elaborates on the tension between repairing the world in acts of *tikkun olam* and acts that claim that the world is ours to control utterly.

Given all this caution, what if halachic considerations lead us toward supporting a ban on genetic research on human embryos? What would this mean for public policy? What would be lost and gained? By the same token, what if we understood the Jewish position as mandating this research in an uncertain political climate? Would our stand imply an activist role for our leadership? Does a general obligation to heal include all possible avenues, and are we obligated even if the consideration of justice would mandate other research be pursued? In other words, it is not enough for us to consider the question theoretically. If the work is mandated healing, the correct Jewish moral role would be consistently to insist on and advocate for such a position, for to do any less might well be a neglect of a commanded act. In so doing, we must recall that this action of mandated healing is surely not the only place for commanded acts toward the health of our fellow humans.

For Jews, the context of all genetic research touches on sensitive issues of eugenics. At each juncture, one must ask, what of malevolent use of genetic research? The *Shoah* (Holocaust) changed the entire landscape of medical research. Although not only Jews have reason to raise deep concern about the evil specter of genetics, Jews certainly must do so as a primary consideration. Our firmness in remembering history and our disciplined stand to avoid any chance of repetition cannot overcome all efforts at new genetic research. In many ways, the state's horrific use of genetic technology seem less a hazard than the temptations of medicine itself. The link among somatic improvement, class stand, and subsequent power has been made in other work. But critical issues, such as the meaning of difference and aging responsibility of a whole society to bear the vulnerability that illness and disability carry, will be raised by possibilities inherent in this technology. How will the dynamics of power drive, for example, research funding for interventions?

Ends cannot be controlled without close regulation and enforcement of research. It is difficult but not impossible to imagine how to make this feasible. In this technology, one is not intending to create new persons, only new personal parts. Yet, all genetic alteration is surrounded by public fear of such alteration, marginalization, and use of unwarranted power in the hands of the malevolent. Is this the first step in an unacceptable alteration of human species by genetic means? It is possible, of course, but counter to this fear we must raise the understanding that not all genetic research will lead inevitably to the worst possible state excesses. As ethicists, it will be a key part of our shared discourse not only to worry about possible abuses of power, but also to raise concerns about unwarranted fears that might unduly block research efforts.

What of marketplace pressures on this technology? If state-supported evil seems an unlikely telos, the drive for profit might provide an alternative source of concern for maleficence, and here Jewish business ethics offers correctives. The field of ART was marked by its unrestrained use of the marketplace. Without careful oversight, lengths that are permitted for individuals to pursue are unlimited. With new technology that will powerfully extend human life and potentially alter moral meaning necessarily, can we finally offer ethical guidelines to inform policy? Rabbinic prohibitions on unfair marketplace exchanges, or limits on the instrumental use of the body of another, can be used to regulate this arena. I maintain rabbinic norms that are found in sources removed from medical consideration, but related to civil law and justice, might be mobilized to assist our thinking about the just use of technology. The next steps are a call for research and discourse within the Jewish tradition.

These truly historic changes in science will reconstruct not only medicine, but also the basic view of the self. Scientific development we are currently facing calls for an imperative and deeply informed discussion within a time period that is responsive to the rapidity with which new advances in genetic research emerge.

Scholars of religion, theologians, and bioethicists have been asked to reflect carefully on the breathtaking and sweeping changes in medicine and research science. But our role, if prudently undertaken, cannot be accepted without a thoughtful and contextual account of the field of genetics as a whole. We will have to ask tough questions about how a specific technology will relate to other pieces of research, such as reproduction technology, nuclear genetic transfer, and inheritable human genetic manipulation (germ line intervention). We must have courage to resist a rush toward a swiftly moving future, courage to believe that ethical and justice considerations must be taken into account at all stages of research, and moral imagination to see beyond the perimeters of what we are given to what we might do and who we ought become.

The Jewish textual tradition insists that the whole of the intellectual proposition of ethics is linked both to practicality and to prophesy, which means that one's epistemology must be sound and one's vision intact. Judaism insists that what is given, and what is now a fixity, can be changed and imagined beyond. It is the act of moral imagination that this research calls us to make. But the leap from the present to the possible future will take certain conditions.[10]

First is the passion for just citizenship, for the idea that broad social liberation must take place in a responding and listening community. Next is consideration for the vulnerable stranger. Finally, Jewish thought reminds us that the world we stand in now is ours only as stewards, and we will have to reflect carefully beyond the rhetorical flourish of that phrase to core issues of regulation and tough standards of enforcement. We can set limits on research only when we can ensure a large public and plural discourse in which the need for public justice, passion for science, claims of patients, the call of civil dissonance from other religious voices, and competing needs of the marketplace will contend for our attention.

In our first, careful thinking about this new technology, and in our sober reflections and our tendency toward caution—which I maintain is a good and prudent response—we should not be blinded to the extraordinary event of this discovery. This is a stunningly important moment in the history of medicine, one with potential to save and sustain human life. The work that I have seen, cardiac cells beating steadily in the laboratory, nerve cells spinning out their tendrils, is impressive and bold, and challenges us to imagine beyond what is into what is possible. It challenges our moral sensibilities and our moral imaginations. It reminds us of a special blessing that is said when one sees a wise secular scholar pass by, in praise of a Creator who makes human wisdom tangible: "Blessed are You, Ruler of the Universe, who has given of Your knowledge to human beings."

In our cautionary deliberations of telos, process, meaning, and justice, we will have to place in the foreground the essential ethicist's question of whether this is a right act and what makes it so, of how this act can repair a broken world, or of whether it might not find a place in a world so broken, but we cannot forget our responsibility to support the extraordinary gesture of research science that such a discovery represents.[11]

## Notes

1. Jewish law, unlike American secular law, describes four categories of possible actions that are based on the relationship among morality, *halachah* norms, and laws of the secular nation-state. An action may be permitted, or at least unpunishable under the halachic code, but morally undesirable; an action may be permitted and desirable; an action may be prohibited (even if desirable); and an action may be permitted by Jewish law, but prohibited by the secular state and thus not be permitted in Jewish law, since "the law is the law of the land (*dinah d'malchuta dinah*).

2. Questions moderns consider important will be absent from the rabbinic and responsa literature. The primary value placed on community, considerations of justice, obligations to the poor and to strangers, and sexuality and procreativity enthusiastically promoted are the subject of the longer essay from which this is excerpted.

3. Since this research has not been the focus of medical issues that have arisen for patients, and since Jewish law is case driven (no cases, no responsa), the literature is thin. The intent of this work is to direct specific attention to this emerging issue and to stimulate serious inquiry in this direction.

4. It raises the very interesting concept of persons as "text" with multiple "embedded narratives." Note how we conceptualize the human person in a postmodern way: a text with the potential for alternative narratives.

5. See Shannon, in Peters (1998). In this formulation, one can differentiate the embryo at the time before and the period after the appearance of the primitive streak, a line of division in the embryo that is the first step in the formation of a spinal cord of one individual.

6. This line of reasoning creates substantive problems in regulation. In what sense is that subject an entity to which one can claim patent rights?

7. The Greek translation assumes the exact opposite. The word in question is *ason*, which Hebrew translates as "harm," but the Greek renders the word as "form" yielding something like "if there yet be no form, he shall be fined, but if there be form, shalt thou give life for life." The "life for life" clause was thus applied to the fetus instead of the mother.

8. This account is most likely to have been speculative and not real history.

9. In the story of Rachel and Leah, mandrake plants were used to enhance fertility (Genesis 30:14).

10. Rabbinic reasoning works by analogy. In thinking about any new case, such as the invention of electricity, exploration of America, or use of anesthesia in surgery, rabbinic authorities had to seek parallel cases that offered precedent. In this case, framing the analogous case will be of central importance. For example, is the development of hES cell technology more like cloning or more like transplantation?

11. This final section is taken from Laurie Zoloth-Dorfman (1998).

## References

Bliech, D. 1981. *Judian and Healing: Halakhic Perspectives*. New York: KTAV.

Bliech, D. 1999.

Dorff, E. N. 1999. *Matters of Life and Death*. Philadelphia: Jewish Publications Society.

Feldman, D. 1968. *Birth Control and Jewish Law*. New York: New York University Press.

Feldman, D. 1995. *Birth Control and Jewish Law*, 3rd ed. New York: New York University Press.

Jacobowitz, I. 1959. *Jewish Medical Ethics*. London: Bloc.

*Jerusalem Report*, April 16, 1998.

Levinas, E. 1990. *Nine Talmudic Readings*. A. Avonowich, trans. Bloomington: Indiana University Press.

*Midrash*. 1957. Rabbah, Bereshit. Soncino Press, 1957.

Pellegrino, Ed. 2000. *NBAC Testimony, Religious Perspectives. Babylonian Talmud* Pesahim 110b.

Shannon, T. 1998. Genetics, ethics, and theology: The Roman Catholic discussion. In *Genetics: Issues of Social Justice* (Peters, T., ed). Cleveland, OH: Pilgrim Press.

Warren, M. A. 1998. *Moral Status: Obligations to Persons and Other Living Things*. Oxford: Oxford University Press.

Zoloth-Dorfman, L. 1998. Mapping the normal human Self. In T. Peters, ed. *Genetics*. Cleveland, OH: Pilgrim Press.

Zoloth, L. 2000. The ethics of the eighth day: Jewish perspective on human embryonic stem cells. *Report to NBAC*.

# 10

# Roman Catholic Views on Research Involving Human Embryonic Stem Cells[1]

Margaret A. Farley

The Roman Catholic moral tradition offers potentially significant perspectives on questions surrounding research on human embryonic stem cells. I use the plural, "perspectives," because the Catholic community has no uncomplex, single voice on such questions. There is, however, a shared community of discourse, so that one can easily identify common convictions expressed in a common language as well as specifically divergent views on this and other particular moral issues.

First, the common convictions. The Roman Catholic tradition is undivided in its affirmation both of the goodness of creation and the importance of human agency in its continuing processes. God is actively present in the world, and humans are called to discern the sacredness of creation and their own responsibilities as, in a sense, co-creators with God. With one mind Catholics affirm also the importance of both the individual and the community, seeing these not finally as competitors but as essentially in need of each other for the fulfillment of both. It is never possible therefore to justify, in an ultimate sense, the sacrifice of an individual to the community, or to forget the common good when thinking about the individual. It is also clear that humans are responsible for their offspring in ways particular to humans, and that future generations matter both in this world and in a hoped-for unlimited future. This implies that a goal of longer and longer life spans is not an unqualified or in itself absolute good. This has some relevance for arguments for stem cell research that suggest a major goal of a greatly expanded human life span.

The Catholic tradition is unified in its belief in God's active and intimate care for the world and each person in it, and in our own correlative obligations to care for those who are in need, preventing unjustified harm, alleviating pain, protecting and nourishing the well-being of individuals and the wider society. The tradition has deep roots that anchor a commitment to the most poor, the most marginalized, and the most ill, and doing so sustains a commitment to human equality in its most basic sense.

At the same time, some Catholics (whether moral theologians, church leaders, or general members of the Catholic community) clearly disagree on, for example, particular issues of fetal and embryo research, assisted reproductive technologies, and prospects for morally justifiable human stem cell research. These disagreements include conflicting assessments of the moral status of the human embryo and use of aborted fetuses as sources of stem cells.

So much agreement on fundamental approaches to human morality, yet disagreement on specific moral rules, is not surprising. For one thing, affirmations of the goodness of creation, human agency, and principles of justice and care do not always yield directly deducible recommendations on specific questions such as stem cell research. Or again, genuine concerns for the moral fabric of society do not by themselves settle empirical questions regarding possible good or bad consequences of particular technologies. There is, for example, often no easy and direct way to determine whether a particular set of choices regarding scientific research will violate the rights of some persons to basic medical care or undermine respect for the dignity of each individual.

At the heart of tradition, however, is a conviction that creation is itself revelatory, and knowledge of the requirements of respect for created beings is accessible at least in part to human reason. This is what is at stake in the tradition's understanding of natural law. For most of its history, Catholic natural law theory has not assumed that morality can simply be "read off" of nature, not even with the important help of Scripture. Nonetheless, what natural law theory does is tell us where to look; that is, to the concrete reality of the world around us, the basic needs and possibilities of human persons in relation to one another, and to the world as a whole. Looking (to concrete reality) means a complex process of discernment and deliberation, a structuring of insights, a determination of meaning from the fullest vantage point available, given a particular history that includes the illumination of Scripture and accumulated wisdom of the tradition. Hence, the intelligibility of "realities" is not such that their meaning is immediately obvious. What is given to our understanding through experience is not only always partial, but it must always be interpreted. The limits, yet necessity, of this process account for many disagreements about specific matters, even within the faith community.

This brings us to disagreements regarding human embryonic stem cell research. Those who stand within the Catholic tradition tend to look to the reality of stem cells and, what is more relevant in this instance, to the realities of the sources of cells for current research: human embryos and fetuses. A case can be made both against and for such research, each dependent on different interpretations of the moral status of the embryo and the aborted fetus. First, significant numbers of Catholics, including present spokespersons for the American bishops, make the case *against* (Doerflinger 1999; Donum Vitae 1987; Grisez 1990). They hold that human embryos must be protected on a par with human persons, at least to the extent that they ought not to be either created or destroyed merely for research purposes. Moreover, the use of aborted fetuses as a source of stem cells, although not in one sense different from harvesting tissue from any human cadavers, nonetheless should be prohibited as it is complicit with and offers a possible incentive for elective abortion. (If the fetuses in question were spontaneously aborted, however, some opening is allowed for their use in this research.[2]) Part of the case against embryo stem cell research also rests on identifying alternatives (adult cells, dedifferentiated and redifferentiated into specific lineages[3]). One can also presume that the case against embryo stem cell research includes a case against cloning, if and insofar as this research incorporates steps involved in procedures for cloning, such as somatic cell nuclear transfer.[4]

On the other hand, a case for human embryo stem cell research can be made on the basis of positions developed within the Roman Catholic tradition. Growing numbers of Catholic moral theologians, for example, do not consider the human embryo in its earliest stages (before development of the primitive streak or implantation) to constitute an individualized human entity with the settled inherent potential to become a human being.[5] In this view the moral status of the embryo is therefore not that of a person, and its use for certain kinds of research can be justified. Since it is, however, a form of human life, some respect is due it; for example, it should not be bought and sold. Those who make this case prefer a return to the centuries-old Catholic position that a certain amount of development is necessary in order for a conceptus to warrant personal status.[6] Embryologic studies now show that fertilization (conception) is itself a process (not a moment), and provide warrant for the opinion that in its earliest stages (including the blastocyst stage, when the inner cell mass is isolated to derive stem cells for purposes of research) the embryo is not sufficiently individualized to bear the moral weight of personhood. Moreover, some concerns regarding aborted fetuses as sources for stem cells can be alleviated if safeguards (such as ruling out direct donation[7] for this purpose) are put in place, not unlike restrictions articulated for general use of fetal tissue for therapeutic transplantation. Finally, concerns about cloning may be addressed at least partially by insisting on an absolute barrier between cloning for research and therapeutic purposes on the one hand, and cloning for reproductive purposes on the other. The latter, of course, raises many more serious ethical questions than the former.

We have, then, two opposing cases articulated within the Roman Catholic tradition. It would be a mistake to conclude that what this tradition has to offer, however, is only a kind of draw. It offers, rather, a continuing process of discernment that remains faithful to a larger set of theological and ethical convictions that takes account of the best that science can tell us about some aspects of reality, and that aims to make one or the other case persuasive on the basis of reasons whose intelligibility is open to the scrutiny of all. I myself stand with the case for embryonic stem cell research, and I believe this case can be made persuasively both within the Catholic tradition and in the public forum. The newest information we have from embryologic studies supports this case, and I believe that it can be made without sacrificing the tradition's commitments to respect human life, promote human well-being, and honor the sacred in created realities. Furthermore, to move forward with this research need not soften the tradition's concerns to oppose the commercialization of human life and to promote distributive justice in the provision of medical care.[8]

Our tradition's conversation on such matters yields more light than I have space to show. It is also a reminder to all of us of the importance of epistemic humility, especially if and as we decide to open more and more room for the human control of creation.

## Notes

1. This essay is adapted from Dr. Farley's testimony before NBAC, May 7, 1999.

2. The difficulty often noted regarding this option, however, is that spontaneously aborted fetuses are frequently not a source for healthy cells or tissue (there is a reason why they spontaneously aborted).

3. See, for example, Pittenger et al. (1999) and Wade (1999). This alternative could prove to be extremely important precisely because it does not involve harvesting stem cells either from embryos or from aborted fetuses. Many scientists, however, consider this alternative as too far away, in terms of research still necessary to develop it, to be a realistic competing possibility.

4. There is not space to expand on the relevance of this point. But some stem cell research, at least, does involve the first stages of cloning, although the goal is not to bring a clone to birth. See the chapter by Erik Parens in this volume.

5. See, for example, Donceel (1970). Early views on this matter were, of course, based on inadequate knowledge of reproductive biology; and twentieth-century views that hold the presence of potential for personhood from the moment of conception are based on more adequate knowledge. The contemporary position on delayed hominization, however, is argued on the basis of more recent embryologic studies. For the Catholic tradition, science is extremely important for theology, although not in every case determinative.

6. See, for example, Shannon and Walter (1990), McCormick (1994), and Cahill (1993).

7. That is, ruling out the possibility of a woman who elects abortion and directly donates fetal stem cells for therapeutic treatment of someone she knows. Other safeguards insist that the investigator not be the attending physician for an abortion.

8. These and other concerns are urgent with regard to the overall question of human stem cell research. However, there is not space to pursue them, or even articulate them, here.

## References

Cahill, L. S. 1993. The embryo and the fetus: New moral contexts. *Theological Studies* 54: 124–142.

Doerflinger, R. 1999. Destructive stem-cell research on human embryos. *Origins* 28: 769–773.

Donceel, J. 1970. Immediate and delayed hominization. *Theological Studies* 31: 76–105.

*Donum Vitae (Respect for Human Life)*. 1987. *Origins* 16: 697–711.

Grisez, G. 1990. When do people begin? *Proceedings of the American Philosophical Association* 63: 27–47.

McCormick, R. A. 1994. Who or what is the preembryo? In *Corrective Vision: Explorations in Moral Theology*. Kansas City: Sheed & Ward.

Pittenger, M. F., Mackay, A. M., Beck, S. C., Jaiswal, R. K., Mosca, D. R., Moorman, M. S., Simonetti, D. W., and Marshak, C. S. 1999. Multilineage potential of adult human mesenchymal stem cells. *Science* 284: 143–147.

Shannon, T. A. and Walter, A. B. 1990. Reflections on the moral status of the preembryo. *Theological Studies* 51: 603–626.

Wade, N. 1999. Discovery bolsters a hope for regeneration. *New York Times*, April 2.

# 11

## Human Embryonic Stem Cells: Possible Approaches from a Catholic Perspective

Michael M. Mendiola

My comments are directed to the general issue of religious ethics and ethical assessment of human embryonic stem (hES) cell research. However, as there is no generic religious ethics, I focus these comments in a very particular way; that is, in response to the Roman Catholic-Christian tradition and concerns and objections raised within it regarding the ethical permissibility of this research. My goal is not to challenge or assess critically objections raised within this tradition, but to suggest two possible approaches within the tradition that may allow Catholic theologians not necessarily to approve ethically such research, but at the minimum to achieve an ethical modus vivendi with it. I am motivated equally by respect for the deeply held convictions and reasoned judgments of my Catholic colleagues, as much as by a desire to move, if possible, beyond the kind of impasse we see regarding an equally contentious matter—abortion. Hence, I do not suggest the approaches so much as corrections to the tradition as an invitation to my Catholic colleagues to begin a critical dialogue about them from within the tradition.

In my judgment (and in the judgment of others[1]) one of the central Catholic objections to this research lies in the source of hES cells: embryos and their destruction and/or use. This source raises dual concerns: the tradition's prohibition of deliberate killing of persons (with the internal stipulation that embryos are persons) and cooperation or complicity in moral wrongdoing. It is the destruction of embryos that poses the greatest challenge or barrier from this tradition's perspective.

The first avenue of investigation concerns the methodology used in ethical analysis of stem cell research. The inviolability of human life is sometimes used in an absolutist sense in Catholic analyses of this research, somewhat like a trump card that stops discussion and leads to a seemingly insurmountable impasse in public discourse. This kind of absolutist use of moral principles can be challenged by the tradition itself. Within the tradition lies a broad and sophisticated stream of ethical reflection on social and political life known as Catholic social teaching. Within social teaching, moral principles consistent with the tradition's dual sources of reason and faith, such as the dignity of the human person, were developed and articulated relative to various social and political spheres. What is crucial to note, however, is that the manner of their application to concrete conditions has generally been perceived as fluid, dynamic, and historically and contextually conditioned. Throughout social teaching the need to read signs of the times, so to speak, as an essential aspect of social ethical analysis is affirmed. Thus the church's teaching must be to shape but also be shaped by the social realities and context of the teaching.

For example, the Second Vatican Council, a ground-breaking council of the whole church in the 1960s, in its historic pastoral constitution on the church (*Gaudium et Spes* 1996), distinguished between its central, normative teachings and their application to concrete circumstances. The latter always involve "the changeable circumstances which the subject matter, by its very nature, involves" (*Gaudium et Spes* 1996, 244). The council thus explicitly held that some of its more concrete proposals were provisional, given the mutability of historical circumstances. It even noted that Roman Catholics may differ and disagree regarding the exigencies of concrete situations and proposed solutions to address them (*Gaudium et Spes* 1996). Leslie Griffin (1990, 336) offered this point more generally:

in Catholic social teaching (in contrast to Catholic sexual ethical thought) the hierarchical magisterium has formulated a general framework of norms and principles for social life, while at the same time respecting political, economic, historical and cultural differences among people of every nation. The church has been careful not to offer solutions of "universal validity"[2] ... which disregard individual differences. Nor has it wished to usurp areas of lay competence (the technical aspects of economics, political science, etc.); instead, the laity are encouraged to apply the church's teachings to their specific circumstances. These features result in a style of social ethics which is inductive, able to learn from concrete experience in circumscribed settings.

The methodologic implications of this contextual-specific recognition relative to the church's teaching are important to note. At a minimum, it suggests that an ethical position or judgment must be based not simply on generalized principles, but also on full attention to the historical and social matrix within which the issue at stake arises. Furthermore, it suggests that a moral principle gains its full intelligibility only in interaction with the concrete context regarding which it is brought to bear.

When applied to hES cell research, this social ethical methodology requires serious consideration of the current social, scientific, and cultural context of the research when invoking a norm such as prohibition of destruction of human or personal life. It requires attention at least to the following: that such research is already being undertaken; that its therapeutic potential seems great; that concrete, suffering human beings may benefit greatly from future therapeutic applications; that people of good will differ regarding the question of the moral status and inviolability of embryonic life; and that people of good will differ regarding the moral permissibility of hES cell research itself. In light of such considerations, one might even make the case that the norm has wider application than simply to embryos. To the degree that the future is influenced in some part by what we do now, the future lives of real, concrete human beings may indeed be harmed or destroyed by our unwillingness in the present to undertake such research. In other words, inclusion of historical-social features of the current context may widen or, at the very least, make more complex the meaning and range of applicability of the principle prohibiting destruction of human or personal life. Richard McCormick (1985, 139) summarized my point well: "... Catholics must learn to distinguish between universally binding moral principles and specific applications. The latter allow for diversity of opinion.... This is *a fortiori* true of political choices. To fail to make this distinction is to degrade teaching authority." Whereas I am less concerned than McCormick about degrading teaching authority, I am deeply concerned about the tenor and quality of Christian ethical reflection in this regard.

Let me stress again that my efforts here are only meant as suggestive, offered with full cognizance that others will disagree substantially with them. My point, however, is modest: it is to invite Catholic moral theologians and scholars to consider and to explore the implications of adopting modes of moral reflection evident in social teaching as an appropriate manner in which to frame its ethics of hES cell research, and one faithful to its own heritage of ethical reflection.

My second approach lies in modification of what others have termed an ethic of compromise. An ethic of compromise suggests that we must often make difficult choices between competing values and goods in social and political life, doing, as Charles Curran stated, the best we can in the face of limited and sinful situations (Griffin 1990, 334). Whereas I agree with the logic underlying this approach, I am uncomfortable with the language of compromise, for it seems to intimate too easily that we may ethically give up or water down our most deeply held convictions. My point, rather, is that we may indeed hold on to those convictions, yet still allow public policies and practices that go against those convictions on good ethical grounds. For this reason, I prefer to name this approach an ethic of toleration rather than of compromise, drawing on the notion of toleration operative in Catholic moral theology.

Catholic moral theology has traditionally recognized in social ethics that one can tolerate an evil in order to avoid a greater evil or to bring about a greater good (Curran 1984, 135). This is not purely a consequentialist stance but entails a number of ethical qualifications. For example, one may never directly will or intend moral evil. Hence one could not licitly tolerate intending or doing moral evil oneself. Toleration also requires what Catholic theologians refer to as "proportionate reason." From this vantage point, no act, viewed in and of itself, is morally wrong, without consideration of the intention of the agent and his or her reasons for acting. "[An] action becomes morally wrong when, all things considered, there is no proportionate reason justifying it" (McCormick 1989, 134). This approach has been highly debated within Catholic circles. It has, however, been advocated by leading Catholic moral theologians. McCormick specified further what proportionate reason entails: "(a) a value at least equal to that sacrificed is at stake; (b) there is no less harmful way of protecting the value here and now; (c) the manner of its protection here and now will not undermine it in the long run" (Griffin 1990, 349). Thus, internal to the standard of toleration are ethical stipulations that must be met. I am not suggesting at this point that hES cell research meets these stipulations. I merely seek to lay out initially what toleration itself entails ethically.

An example of what I have in mind may be useful. In the early 1980s the National Conference of Catholic Bishops (NCCB 1983) undertook a long process of ethical deliberation on nuclear warfare, which resulted in their pastoral letter *The Challenge of Peace*. In conducting their deliberations, the bishops held hearings across the United States, inviting a wide range of experts from a number of fields to assist them. That they sought a wide variety of viewpoints embodies and is consistent with the kind of social ethical approach I stated above should characterize Catholic moral reflection. At one point in their deliberations they struggled with the ethical assessment of nuclear deterrence, having already condemned the use of nuclear arms. Their position on deterrence is instructive: although the bishops saw it as morally problematic, they were willing to allow it provisionally and conditionally. In doing so, they drew explicitly on the words of Pope John Paul II: "In current conditions 'deterrence' based on balance, certainly not as an end in itself but as a step on the way toward a progressive disarmament, may still be judged morally acceptable. Nonetheless in order to ensure peace, it is indispensable not to be satisfied with this minimum which is always susceptible to the real danger of explosion" (NCCB 1983). In other words, for the Pope and the bishops, nuclear deterrence was not perceived as a positive good, but could be tolerated as a lesser of two evils, given that other goods (e.g., protection offered by deterrence) were at stake. In light of McCormick's point, we might say that the bishops perceived a proportionate reason for limited, provisional acceptance of deterrence, provided that deterrence was coupled with efforts to move beyond nuclear armaments altogether.

I believe that these constructs of toleration and proportionate reason may allow Catholic theologians to engage in debate about the ethics of hES cell research in a manner consistent with the principle of the inviolability of embryonic life and yet not exclusively dependent on it as a knock-down, stop-the-discourse argument. I offer this suggestion with full cognizance of the complexity involved. For example, the second provision of proportionate reason as specified by McCormick—that there is no less harmful way of protecting the value at stake—clearly raises a challenge to embryonic stem cell research. In its strongest form, it might prohibit research involving embryo destruction altogether until other avenues of stem cell development have been attempted and failed. In its weaker form it would require that researchers consistently seek to move beyond the use of human embryos (although allowing provisional use currently). As noted above, a related, and equally important matter is this tradition's prohibition of cooperation in moral evil. Might an ethic of toleration involve ethically illicit cooperation in the moral wrong-doing of others? A full, analytic, and normative exploration of an ethics of toleration would have to address this range of moral concerns as well.[3] However, even given these complexities, I raise the question to the Catholic moral theological community: is there merit in an ethics of toleration as briefly outlined here?

I suggested that the social ethical framework evident in social teachings and an ethic of toleration may be effective in Catholic assessments of hES cell research. I welcome constructive conversation, particularly from Catholic colleagues, regarding their adequacy, merit, and fruitfulness.

## Notes

1. Testimonies given before the National Bioethics Advisory Committee by Kevin Wildes, Margaret Farley, and Edmund Pellegrino on May 7, 1999, also point to this central concern.

2. Griffin cites Pope Paul VI (1976).

3. For a helpful discussion of cooperation and the various distinctions and criteria utilized to assess cooperation, see O'Donnell (1957) and Ashley and O'Rourke (1989). Ashley and O'Rourke, provide a general statement of what constitutes legitimate cooperation: "*To achieve a well-formed conscience, one should always judge it unethical to cooperate formally with an immoral act (i.e., directly to intend the evil act itself), but one may sometimes judge it to be an ethical duty to cooperate materially with an immoral act (i.e., only indirectly intending its harmful consequences) when only in this way can a greater harm be prevented, provided (1) that the cooperation is not immediate and (2) that the degree of cooperation and the danger of scandal are taken into account*" (italics in the original).

## References

Ashley, B. M. and O'Rourke, K. D. 1989. *Health Care Ethics: A Theological Analysis*, 3rd ed. St. Louis: Catholic Health Care Association of the United States.

Curran, C. E. 1984. An analysis of the American bishops' pastoral letter on war and peace. In *Critical Concerns in Moral Theology*. Notre Dame, MI: University of Notre Dame Press.

*Gaudium et Spes*. 1966. Pastoral constitution on the Church in the modern world. In S. J. Abbot and M. Walker, eds. *The Documents of Vatican II*. New York: America Press.

Griffin, L. 1990. The church, morality, and public policy. In C. E. Curran, ed. *Moral Theology: Challenges for the Future*. New York: Paulist Press.

McCormick, R. A. 1985. *Health and Medicine in the Catholic Tradition*. New York: Crossroad Publishing.

McCormick, R. A. 1989. Pluralism in moral theology. In *The Critical Calling: Reflections on Moral Dilemmas Since Vatican II*. Washington, DC: Georgetown University Press.

National Conference of Catholic Bishops. 1983. *The Challenge of Peace: God's Promise and Our Response*. Washington, DC: United States Catholic Conference.

O'Donnell, S. J. and Thomas J. 1957. *Morals in Medicine*. Westminster, MD: Newman Press.

Paul IV. 1976. *Octegesima Adveniens*. In J. Gremillion, ed. *The Gospel of Peace and Justice*. Maryknoll, NY: Orbis Press.

# 12

# Embryonic Stem Cells and the Theology of Dignity

Ted Peters

Human embryonic stem cells have become the gold mine of genetic prospecting because of their potential for rejuvenating failing organ tissue. (see Chapters by Okarma and Thomson). Ethical arguments demanding that we shut down such potentially life-saving and life-enhancing research must therefore come from some strong and compelling commitment.

It is my opinion that this strong commitment is to human dignity, and that the orienting question is whether dignity applies to the blastocyst that is destroyed when obtaining pluripotent cells. The dignity of the early embryo has become the central ethical issue in the public debate over the advisability of continuing human embryonic stem (hES) cell research. This orienting question is shared by both proponents and opponents of the research. Opponents often base their stand on the assumption that the blastocyst must be treated as a being with dignity. Proponents try to create a divide between the pluripotent stem cell and the embryo; in so doing, they implicitly also concur that embryos must be treated as beings with dignity. To avoid being accused of violating that dignity, scientists claim that stem cells are only pluripotent and not totipotent, as is the embryo. Hence, pluripotency functions ethically to exonerate the researchers. My own view is that this comes perilously close to subterfuge; and subterfuge is insufficient grounds for ethical exoneration.

I hold that, in principle, every cell in our body might become a potential embryo and, in turn, a potential new person. We can now imagine a future of baby making that may not be limited to fertilization of ova, let alone to sexual procreation. As long as the genetic potential for a new person resides in a given cell, what distinguishes it as a potential human being is its relation to its environment. Becoming a human being requires more than a genome; it requires an intentional, nurturing, relational community. Such a community is missing in laboratory experimentation. This is not a criticism and not a concession. It is an acknowledgement of an ethically relevant difference.

I will examine the concept of dignity, with special interest in a theological understanding of dignity. I work with the Kantian assumption regarding personhood, according to which a person should always be treated as an end and not merely as a means to some end. Even though the prevailing view is that such dignity is innate, perhaps even datable to an embryo's fertilization, I contend phenomenologically that dignity is conferred. Dignity is a relational concept that begins first with the external conferral of dignity before it is claimed by a person as something intrinsic.

This foray into theological anthropology will emphasize the Christian commitment to the human person as a whole. Holism is inclusive of body and soul, inclusive of genes and relationships. In addition, it is inclusive of a person's entire life story. Essential to the human reality is resurrection, the eschatologic fulfillment of each of our life stories in God's new creation. Rather than locate human dignity at the point of origin, I suggest that it is our destiny that is theologically decisive.

Finally, I return to the central ethical question and ask whether understanding dignity as eschatologically conferred illumines the debate over stem cell research. This will not answer the question decisively, unfortunately, even though it provides considerable ethical support to molecular biologists pursuing stem cell research.

## The Ethical Question—A Shared Premise

Whereas hES cell research presents a host of ethical questions, the question of the dignity of the embryo seems to tap the most energy. Frank E. Young, former Commissioner of the U.S. Food and Drug Administration, writes, "The devaluation of humans at the very commencement of life encourages a policy of sacrificing the vulnerable that could ultimately put other humans at risk, such as those with disabilities and the aged, through a new eugenics of euthanasia" (2000). Note the gravity of the concern expressed by this rhetoric, according to which stem cell research is "sacrificing the vulnerable" leading to "euthanasia."

An organization called Do No Harm: The Coalition of Americans for Research Ethics lobbies against U.S. government-funded research that destroys human embryos:

> That some individuals would be destroyed in the name of medical science constitutes a threat to us all. Recent statements claiming that human embryonic stem cell research is too promising to be slowed or prohibited underscore the sort of utopianism and hubris that could blind us to the truth of what we are doing and the harm we could cause to ourselves and others. Human embryos are not mere biological tissues or clusters of cells; they are the tiniest of human beings. Thus, we have a moral responsibility not to deliberately harm them.... The last century and a half has been marred by numerous atrocities against vulnerable human beings in the name of progress and medical benefit. In the 19th century, vulnerable human beings were bought and sold in the town square as slaves and bred as though they were animals. In this century, the vulnerable were executed mercilessly and subjected to demeaning experimentation at Dachau and Auschwitz.... These experiments were driven by a crass utilitarian ethos which results in the creation of a sub-class of human beings, allowing the rights of the few to be sacrificed for the sake of potential benefit to the many. (Center for Bioethics 1999)

The stated premise is that each human embryo is the "tiniest of human beings." The unstated second premise is that, because a zygote is already a tiny human being, it has dignity. Having dignity, it deserves protection from scientists who would destroy it in the name of medical research. The "do no harm" medical maxim applies here and is violated in hES cell research.

This notion suggests that stem cell research is driven by a crass utilitarian ethic that sacrifices the dignity of individuals by turning them into a means toward an end. By implication, if we adopt this crass utilitarianism, we risk repeating atrocities such as slavery and concentration camps. Hence, Do No Harm rejects stem cell research, even if it leads to a laudable end such as medical advance. "We are simply not free to pursue good ends via unethical means. Of all human beings, embryos are the most defenseless against abuse" (Center for Bioethics 1999). Do No Harm relies on the Kantian concept of dignity as "an intrinsic unconditioned, incomparable worth" (Kant 1948), which means that a human person is an end and may never be used merely as means to a further end.

Does the early embryo have dignity and, if so, does this provide sufficient warrant to prevent its destruction for purposes of research? Do No Harm would join many Roman Catholic bioethicists (Clarke 2000) in answering yes, the embryonic stem cell does have dignity; therefore, such research is immoral and should be prohibited by law.

I submit that this same ethical assumption is silently assumed by those who advocate the research. The American Association for the Advancement of Science (AAAS), for example, maintains that "Isolated from the total structure of the embryo or blastocyst, these cells, even under favorable conditions, will not develop the trophoblast (the outer layer of cells of the embryo) or other structures needed for continued development. Another way of putting this is to say that stem cells are pluripotent rather than totipotent" (Chapman et al. 1999). The scientific distinction between totipotency and pluripotency carries ethical weight: if cells are totipotent they have dignity and may not be destroyed; if they are merely pluripotent they do not have the same dignity and can be destroyed. The tacit assumption is that totipotency is ascribed only to the zygote and early embryo. The logic is that if we can limit public policy discussion to pluripotent stem cells we can avoid acknowledging the dignity question implicit in the concept of totipotency. The objective of research advocates is to remove pluripotent stem cells from the category of embryos and thereby exempt them from ethical opprobrium and from legal restrictions on research.

In short, advocates operate on the assumption that there is a hierarchy of cells. In that hierarchy, hES cells are pluripotent, not totipotent. Hence, although advocates support hES cell research, they do not necessarily operate out of a completely different set of ethical assumptions than those that undergird opposition to the research.

## Challenging the Ethical Premises

Much seems to depend on whether totipotency does indeed confer dignity, and whether pluripotent stem cells such as hES cells can be said not to have this same dignity. In its *Guidelines for Research Involving Human Pluripotent Stem Cells*, for example, the National Institutes of Health states bluntly: "Although human pluripotent stem cells may be derived from embryos or fetal tissue, such stem cells are not themselves embryos" (U.S. Department of Health and Human Services 2000). Two assumptions are therefore undergirding current arguments: first, that a sharp distinction can be drawn between pluripotent and totipotent cells, and second, that dignity is something that inheres in a totipotent cell and in that cell alone. I challenge both premises.

Can a sharp distinction be drawn between pluripotent and totipotent cells? At first glance, it appears that it can. Certainly, much of the scientific debate has assumed so. But consider the following (Strauss 1999):

Research into tissue specific stem cells is yielding promising surprises. Their potential for renewing may be transferable. Experiments with mice have successfully transferred neural stems cells from the brain to the bone marrow, resulting in the production of blood. Once transplanted from the brain into the bone marrow, the neural stem cells produced a variety of blood cell types including myeloid and lymphoid cells as well as early hematopoietic cells. This seems important in two ways. First, the neural stem cells appear to have a wider differentiation potential than what is required to produce brain tissue (Bjornson et al. 1999). Second, some kind of triggering mechanism must be present in the blood system that can instruct the stem cell genes to produce blood. Anticipating future medical value, this brightens the prospect that neural cell transplants might be able to treat human blood cell disorders such as aplastic anemia and severe combined immunodeficiency.[1]

The plasticity of stem cells that were formerly presumed to create only one type of differentiated daughter cell has important implications for thinking about potency. "Evidence is mounting that the findings are not aberrations but may signal the unexpected power of adult stem cells" (Vogel 2000, 1419; c.f. Clarke et al. 2000, 1660). What is at stake is the reprogramming of cells so that they can become many types of tissues. We have to determine just how cytoplasm interacts with the DNA nucleus and be able to reprogram cytoplasm to make specific tissue. Once this ability to reprogram is achieved, in principle it could apply to any cell. We would not necessarily at that point have to rely on oocytes or fertilized ova or, perhaps, even blastocysts as the source. Somatic cells might become the source for pluripotent cells.

Might we begin to think of each cell in our body as an embryo? Would this mean that, in principle, we could make a baby from any cell in our body? Here is all that we need: the full genetic code to make every tissue available in every somatic cell; the ability to return our DNA nucleus to quiescence and then to its predifferentiated state, as in the case of Dolly; and the ability to reprogram cytoplasm to cause selected genetic expression and, together with this, to initiate embryonic development. This is all it takes. The first two are already established. Nature has given us a full complement of genes in every somatic cell. Cloning experiments at the Roslin Institute gave us the technology of quiescence for returning an already differentiated somatic nucleus to its predifferentiated state and, hence, totipotency. Only the third scientific task remains to be accomplished. This would demonstrate the principle that babies can come from anywhere.

This may be speculative today, but today's speculation may be tomorrow's challenge. We would be challenged by a question such as whether every cell in our existing body has the same moral status as a pluripotent hES cell. Or, the same status as a totipotent fertilized ovum.

Thus, whereas current scientific wisdom is that hES cells are pluripotent but not totipotent, the AAAS acknowledges that "advanced technology might be able to render these cells effectively (if not actually) totipotent" (Chapman 1999). The hES cell contains all the same genetic material as the early embryo from which it is derived. It lacks only the environmental structure (trophoblast, etc.) to permit it to become an embryo. Eventually, technology may be able to supplant this lack. If that happens, the presumed difference between pluripotency and totipotency collapses. Does the ethical argument collapse with it?

### Dignity and the Role of the Environment

The AAAS tries to get around the problem by appealing to a distinction between natural and contrived environments: "To fail to distinguish between the natural and contrived development of the embryo would ... unreasonably commit us to the full moral protection of every human cell" (Chapman 1999). The AAAS implicitly acknowledges that if it is totipotency that renders a cell protectable as a human being, every cell, not just the early embryo, might ultimately deserve this protection.

If virtually any somatic cell within our body is a potential human being, what does this do to the status of the early embryo? Does it retain a special status? If the early embryo and every other somatic cell due to laboratory technology gain the same status, the same potential to become a human being, how would the principle of dignity apply? Could we any longer distinguish the presence of dignity for the embryonic stem cell while denying it to pluripotent cells or to the wide array of somatic cells? We might also ask, should we have been applying dignity to cells rather than persons in the first place?

To help us, perhaps we should take a side trip through the concept of dignity. Both sides of the current debate—those opposing destruction of embryos in stem cell research and those defending use of pluripotent stem cells—allow for, if not strongly support, human dignity. The question has to do with where to apply it.

Dignity has to do with the intrinsic value of a human person, and cannot be reduced to his or her instrumental worth. This means that we are always worth more than our stock market portfolios or our reputations or our function in the economy. As persons we dare not be reduced to the subjective value of those who like or dislike us. We know we can claim our rights even when everyone around dislikes us. As individuals we are always an end and never merely a means to some greater value. It is this dimension of intrinsic value that constitutes human dignity as we know it in the modern West.

Let me pose a phenomenologic question: is dignity intrinsic or conferred? It is both. Theologically, I believe our human dignity is ultimately conferred by God. And, I would add, because we have experienced God treating us with dignity, we now confer it on one another. The result of the conferring is that dignity functions for us as intrinsic or even innate worth.

I work with a theological maxim: *God loves each of us regardless of our genetic makeup, and we should do the same.* This is my proposed genetic adaptation of 1 John 4:11: "Beloved, since God loved us so much, we also ought to love one another." One of the ways that we have learned about God's conferral of dignity on us is through the ministry of the incarnate Son that took him to the most humble of persons in first-century Israel: beggars, lepers, those crippled or blind from birth, and social outcasts such as adulterers and traitorous tax collectors. Jesus took special interest in those who suffered marginalization, or who just plain suffered. He was particularly concerned about children. "Let the little children come to me, and do not stop them, for it is to such as these that the kingdom of heaven belongs" (Matt. 19:14; Peters 1996, 52–54; 1998, 116–129). For Jesus, conferring dignity was an ethical activity.

Love creates dignity in the humble. To be the object of someone's love is to be made to feel valuable, to feel worth. Once you or I feel this sense of worth imputed to us by the one who loves us, we may begin to own it. We may begin to claim self-worth. Worth is first imputed, then it is claimed.

In the modern West we typically assume that human dignity is innate. For something to be innate, it must be present at birth. Now that we live in the genetic age, dignity has been pressed backward in time all the way to ovum fertilization and creation of a zygote with the future genome established. This understanding of innate dignity is clearly the position defended by the Vatican (*Donum Vitae* 1987) and apparently by Do No Harm as well.

Such a doctrine permits us in court to defend the rights of every individual regardless of how humble he or she might be. But phenomenologically this view is mistaken. Dignity—at least the sense of it as self-worth—is not simply inborn. Rather, it is the fruit of a relationship, a continuing, loving relationship. A newborn welcomed into the world by a mother and father who provide attention and affection develops self-consciousness that incorporates this attention and affection as evidence of self-worth. As consciousness becomes constituted this sense of worth can be claimed for oneself, and individual dignity develops.[2]

Phenomenologically dignity is relational. Theologically, it is also proleptic; that is, it is fundamentally future oriented. Conferring dignity on someone who does not yet in fact experience or claim it is a gesture of hope, an act that anticipates what we hope will become actuality. Our final dignity, from the point of view of the Christian faith, is eschatologic; it accompanies our fulfillment of the image of God. Rather than something imparted with our genetic code or accompanying us when we are born, dignity is the future end product of God's saving activity that we anticipate socially when we confer dignity on those who do not yet claim it.

Human dignity has to do with God's valuing us, according to Christian faith, especially God's self-emptying love expressed in the incarnation. The late Richard A. McCormick, noted Roman Catholic bioethicist, voiced special concern for the destruction of early embryos. He sought to bring theological resources to bear on ethical questions raised by genetic research. "Since our dignity roots in our origin and our destiny (God), it is clear that we are equally dignified. Bringing such a conviction to the Genome Project and the technologies it generates will not be easy.... Theology provides the essential context for moral reasoning and therefore affects it deeply. Love of and loyalty to Jesus Christ, the perfect man, sensitizes us to the meaning of persons" (McCormick 2000, 426–427). On the other side of the Protestant Reformation we find something parallel. "Reformed confessions implicitly tell us who we are by reminding us of who God is" (Burgess 1998, xiv).

The ethics of God's kingdom in our time and in our place consists of conferring dignity and inviting persons to claim dignity as a prolepsis of its future fulfillment. Conferring dignity confronts enormous challenges. As we look around our world we see dignity denied in almost every quarter. African clan rivalry is producing genocide on a scale of hundreds of thousands. In Latin America repressive political regimes with their death squad terror prevent the exercise of basic rights. Reported child abuse in North America indicates that the most helpless among us are not receiving even basic care and loving attention. Millions of persons the world over are not being treated as having intrinsic value. Actualized dignity is relational, and destructive relations make it into an unrealized ideal rather than an authentic human experience. The biblical mandate to love one another means, among other things, imputing dignity to all persons in such a way that they may rise up and claim self-worth and share in the benefits of living together on this planet. This is our contemporary ethical mandate.

### Dignity and Stem Cell Research

How does this phenomenologic and theological discussion regarding human dignity apply to stem cell research? Beginning with a spirit of charity, I suggest that both sides seek to confer dignity. Those opposing the research seek to confer dignity on the totipotent stem cell, treating it today as if it is already the person it could become tomorrow. Those promoting the research confer dignity on persons who in the future will benefit medically when new therapies are developed. They do not intend to compromise dignity by denying it to pluripotent stem cells. The question is whether they do in fact compromise dignity by denying it to totipotent embryonic cells.

Where might we go at this point? One direction would be to rally with opponents to say our task is to confer dignity on each and every fertilized zygote, and shoulder the responsibility of making certain that it is implanted and brought through pregnancy to birth. As a corollary, we would also say that stem cell research violates the dignity of blastocysts; therefore, we should place a moratorium against it until sources for stem cells other than embryos can be established.

Or, instead we might side with proponents to say, because embryonic stem cells derive from excess fertilized ova at in vitro fertilization clinics, and would never under any circumstances reach implantation or have the natural environment necessary to become a human being, and because hES cells show such enormous potential for developing new therapies that could dramatically enhance human health and well-being, these cells may be used to serve the dignity of future persons who will benefit. Because pluripotent stem cells do not have the actual potential for becoming a human being, they do not have dignity to be compromised. In this case, the dignity honored is that of future beneficiaries of this medical research.

Relevant here, I suggest, is our awareness that the science is fast moving. Whether or not dignity should be assigned to cells in addition to babies is a question that could not have been conceived a half century ago. The problems we face in applying dignity to cells based on their presumed potency could not have been conceived just a few years ago. These problems are likely to give way to new ones if and when cytoplasmic reprogramming becomes a technologic reality. No longer will gametes and fertilized zygotes be the only sources for future human beings. Virtually every cell in our body could be dubbed the "tiniest of human beings." What then will determine whether one cell has actual potential for humanity and another does not?

What will make this determination is the decision made on the part of some living persons to call a somatic cell or a stem cell or a fertilized zygote into a destiny wherein it will become a baby, then a child, then a grownup, then, as Stephen Post (1999) says, a life journey. The potential for a cell to become a person depends on much more than the existence of a full genome; it depends on being called by a future parental relationship.[3]

Whether or not this has always been so is beside the point. The fast-moving frontier of genetic technology will increasingly make this so for the future. This may be a nightmare for those of us who sleep comfortably with the sweet dream that our values are rooted in nature, rooted in things innate, rooted in things inborn. Rather than unwitting and helpless heirs to nature's bequeathal, we are finding ourselves increasingly orphaned by advances in scientific knowledge and genetic technology. Having been thrust from the comforts of natural Eden into the wilderness of technologic anomie, we now have to make choices. The children of the future will be increasingly the result of choices, including choices over very early cells. Nature no longer can make these choices for us. We will be unable to avoid the responsibility of deciding which cells become babies and which do not. And the sheer mathematics of the situation will not permit all cells to become human beings.

Bearing all this in mind, I find myself supporting stem cell research. This does not by any means indicate that I am persuaded by ethical arguments that depend on distinctions between totipotent and pluripotent cells. In fact, I am not finally persuaded at all. I find myself in an interim state, struggling to weigh the complex factors. My theological excursis into dignity is illuminating, but it does not make answering the central ethical question clear enough to be decisive. This may be disappointing to some readers.

In weighing the factors, I find the defense of human dignity by groups such as Do No Harm laudable and noble; yet, on close examination, the defense dribbles away like sand through the fingers when we see how broadly the notion of potential human being can be applied to cells. What becomes compelling in my judgment is the opportunity that appears to be present to advance dramatically the quality of human health and well-being. By no means do I make an appeal to crass utilitarianism here.

Rather, I see the larger enterprise of dedicated scientific research serving the dignity of persons who will tomorrow benefit from difficult laboratory work today.

## Notes

1. An adult neural stem cell has very broad developmental capacity and may potentially be used to generate a variety of cell types for transplantation in different diseases (Clarke et al. 2000, 1660).

2. In emphasizing the role relational community plays in the phenomenon of dignity, I do not want to risk losing the individual in the collective. In the context of opposing human cloning, Jean Bethke Elshtain critiqued the lack of moral structure surrounding the autonomous Western self and identified "two ideological projects that bedevil modern society. Those projects are untrammeled individualism, on the one hand, and collectivism, on the other" (Rantels and Milgram 1999, 151).

3. Openness toward relationships, especially toward a relationship with a transcendent God, is essential to Christian anthropology, and is key to understanding what we mean by the human soul. "Human beings are bodily creatures who have a fundamentally unlimited transcendentality and unlimited openness to being as such in knowledge and freedom" (Rahner 1976–1988, xxi). "A being is the more itself the more it is open, the more it is in relationship" (Auer and Ratzinger 1988, 155).

## References

Auer, J., and Ratzinger, J. 1988. *Dogmatic Theology 9: Eschatology*. Washington, DC: Catholic University Press of America.

Bjornson, C. R., Rietze, R. L., Reynolds, M. C., Magli, A., and Vescovi, L. 1999. Turning brain into blood: A hematopoietic fate adopted by adult neural stem cells in vivo. *Science* 283(5401): 534–536.

Burgess, J. P., ed. 1998. *In Whose Image: Faith, Science, and the New Genetics*. Louisville, KY: Geneva Press.

Center for Bioethics and Human Dignity. 1999. *On Human Ethics and Stem Cell Research: An Appeal for Legally and Ethically Responsible Science and Public Policy*. Available at the Do No Harm web site www.stemcellresearch.org.

Chapman, A. R., Frankel, M. S., and Garfinkel, M. 1999. Stem cell research and applications. Prepared for the American Association for the Advancement of Science and the Institute for Civil Society. Available at www.aaas.org/spp/dspp/sfrl/projects/stem/main.htm.

Clarke, D. L., Johansson, C. B., Wilbertz, J., Veress, B., Nilsson, E., Karlstrom, H., Lendahl, U., and Frisen, J. 2000. Generalized potential of adult neural stem cells. *Science* 288(5471): 1660–1663.

Clarke, K. 2000. Unnatural selection: How biotechnology is redesigning humanity. *U. S. Catholic* 65(1): 14–15.

*Donum Vitae*. (Congregation for the Doctrine of the Faith). 1987. Instruction on respect for human life in its origin and on the dignity of procreation: Replies to certain questions of the day. In T. A. Shannon, ed. *Bioethics: Basic Writings on the Key Ethical Questions that Surround the Major, Modern Biological Possibilities and Problems*, 3ʳᵈ ed. Mahaw, NJ: Paulist Press.

Kant, I. 1948. *Groundwork of the Metaphysic of Morals*. H. J. Patton, trans. New York: Harper.

McCormick, R. A. 2000. Moral theology and the genome project. In P. R. Sloan, ed. *Controlling Our Destinies*. Notre Dame, MI: University of Notre Dame Press.

Peters, T. 1996. *For the Love of Children: Genetic Technology and the Future of the Family*. Louisville, KY: Westminster/John Knox Press.

Peters, T. 1998. Love and dignity: Against children becoming commodities. In R.A. Willer, ed. *Genetic Testing and Screening*. Minneapolis: Kirk House.

Post, S. J. 1999. In G. McGee, ed. *The Human Cloning Debate*. Berkeley: Berkeley Hills Books.

Rahner, K. 1976–1988. *Theological Investigations*. 21 vols. London: Darton, Longman, and Tod. New York: Seabury, 1974–1976. Reprint, New York: Crossroad.

Rantels, M. L., and Milgram, A. J., eds. 1999. Cloning humans is immoral. In *Cloning: For and Against*. Vol. 3. Chicago: Open Court.

Strauss, E. 1999. Brain stem cells show their potential. *Science* 283(5401): 471.

U.S. Department of Health and Human Services. 2000. *Draft National Institutes of Health Guidelines for Research Involving Human Pluripotent Stem Cells*. Waskington, DC.

Vogel, G. 2000. Can old cells learn new tricks? *Science* 287(5457): 1418–1419.

Young, F. E. 2000. A time for restraint. *Science* 287(5457): 1424.

# 13

## Some Protestant Reflections[1]

Gilbert Meilaender

I address the issue of stem cell research in my capacity as a Protestant theologian. At the same time, I cannot claim to speak for Protestants generally. Alas, no one can. I do, however, draw on several theologians who speak from within different strands of Protestantism. A significant number of my coreligionists more or less agree with the points I will make. Others will disagree, even though I like to think that, were they to ponder these matters long enough, they would not.

Moreover, I try not to think of this chapter as an attempt by some Protestant interest group to put its oar into deliberations about stem cell research. Although I will begin as best I can from somewhere rather than nowhere, from within a particular tradition, its theological language seeks to uncover what is universal and human. It begins epistemologically from a particular place, but it opens up ontologically a vision of the human. These reflections may therefore be of interest not only because they articulate the view of a sizeable number of our fellow citizens but also because they seek to uncover a vision of the life we share in common.

I confess that the topic of human embryonic stem cell research raises for me complexities that I do not fully understand. As I tried to follow recent developments, they often seemed bewildering. Nonetheless, perhaps I can bring an angle of vision that will enrich our deliberations.

To that end I will make three points. For each one I will take as my starting point a sentence from a well-known Protestant thinker—not to claim that theologian's authority for or agreement with what I have to say, but to provide some texts with which to begin my reflections.

The first is a passage from Karl Barth, perhaps the greatest of twentieth-century theologians, who wrote from the Reformed (Calvinist) tradition: "No community, whether family, village or state, is really strong if it will not carry its weak and even its very weakest members" (1961, 424).

That sentence invites us to ponder the status of the human embryo, the source of many, although not all, stem cells that would be used in research. One complexity that I do not fully understand involves the question whether stem cells are not themselves and cannot develop into embryos. I assume that they are not and cannot, although perhaps I need to be instructed further on that matter. Even making this assumption, however, we face the fact that procuring embryonic stem cells for research requires destruction of the embryo. Hence, we cannot avoid thinking about its moral status.

No doubt it is, in our society, impossible to contemplate this question without feeling sucked back into the abortion debate, and we may sometimes have the feeling that we cannot consider any other related question without always ending up arguing about abortion. Perhaps there is something to that, and I will not entirely avoid it myself, but the question of using (and destroying) embryos in research is a separate question. The issue of abortion, as it is framed in our society's debate and in Supreme Court decisions, turns chiefly on a conflict between the claims of the fetus and the claims of the pregnant woman. It is precisely that conflict, and our seeming inability to serve the woman's claim without turning directly against the life of the fetus, that is thought to justify abortion. But no such direct conflict of lives is involved in embryo research. Here, as in so many other areas of life, we must struggle to think inclusively rather than exclusively about the human species, about who is one of us, about whose good should count in the common good we seek to fashion. The embryo is, I believe, the weakest and least advantaged of our fellow human beings, and no community is "really strong if it will not carry its ... weakest members."

This is not an understanding shaped chiefly in the fires of recent political debate; rather, it has very deep roots in Christian tradition. To address this issue within that tradition, I must explore those roots briefly. We have become accustomed in recent years to distinguishing between persons and human beings, to thinking about personhood as something added to the existence of a living human being, and to debating where to locate the time when such personhood is added. However, a much older concept of the person, for which no threshold of capacities is required, was deeply influential in Western history and had its roots in some of the most central Christian affirmations. The moral importance of this understanding of the person was noted by Anglican theologian Oliver O'Donovan (1984).

Christians believed that in Jesus of Nazareth divine and human natures were joined in one person, and, of course, they understood that it was not easy to make sense of such a claim. If Jesus had both divine and human natures, he would seem to be two persons, two individuals, identified in terms of two sets of personal capacities or characteristics; a sort of chimera, we might say, in terms appropriate to our subject.

So Christian thinkers turned in a different direction that was very influential in our culture's understanding of what it means to be an individual. In their view, a person is not someone who has a certain set of capacities; a person is simply, as O'Donovan put it, a "someone who"—a someone who has a history. That story, for each of us, begins before we are conscious of it and, for many of us, may continue after we have lost consciousness of it. It is nonetheless our personal history even when we lack awareness of it, even when we lack or have lost certain capacities characteristic of the species. Each story is the story of "someone who," as a living human being, has a history.

This is, as I noted, an insight that grew originally out of intricate Christological debates carried on by thinkers every bit as profound as any we today are likely to encounter. But starting from that very definite point, they opened up for us a vision of the person that carries deep human wisdom, that refuses to think of personhood as requiring certain capacities, and that therefore honors the time and place of each someone who has a history. In honoring the dignity of even the weakest of living human beings—the embryo—we come to appreciate the mystery of the human person and the mystery of our own individuality.

My second text is a sentence from the late John Howard Yoder, a well known Mennonite theologian: "I am less likely to look for a saving solution if I have told myself beforehand that there can be none, or have made advance provision for an easy brutal one" (1974, 91).

Stem cell research is offered as a kind of saving solution, and it is not surprising therefore that we should grasp at it. Although I suspect that promises and possibilities could easily be oversold, none of us should pretend to be indifferent to attempts to relieve or cure heart disease, Parkinson's and Alzheimer's diseases, or diabetes. Suffering and even death are not the greatest evils of human life, but they are surely bad enough, and all honor to those who set their face against such ills and seek to relieve them.

The sentence from Yoder reminds us, however, that we may sometimes have to deny ourselves the handiest means to an undeniably good end. In this case the desired means will surely involve creation of embryos for research and then their destruction. The human will, seeing a desired end, takes control, subjecting to its desire even the living human organism. We must ask ourselves whether this is a road we really want to travel to the very end. Learning to think of human beings as will and freedom alone has been the long and steady project of modernity. At least since Kant, ethics has often turned to the human will as the only source of value. But C. S. Lewis, an Anglican and surely one of the most widely read of twentieth-century Christian thinkers, depicted what happens when we ourselves become the object of this mastering will.

We reduce things to mere Nature *in order that* we may "conquer" them. We are always conquering Nature, because "Nature" is the name for what we have to some extent conquered. The price of conquest is to treat a thing as mere Nature.... The stars do not become Nature till we can weigh and measure them: the soul does not become Nature till we can psycho-analyze her. The wresting of powers *from* Nature is also the surrendering of things *to* Nature. As long as this process stops short of the final stage we may well hold that the gain outweighs the loss. But as soon as we take the final step of reducing our own species to the level of mere Nature, the whole process is stultified, for this time the being who stood to gain and the being who has been sacrificed are one and the same. This is one of the many instances where to carry a principle to what seems its logical conclusion produces absurdity. It is like the famous Irishman who found that a certain kind of stove reduced his fuel bill by half and thence concluded that two stoves of the same kind would enable him to warm his house with no fuel at all.... [I]f man chooses to treat himself as raw material, raw material he will be. (1947, 49–84)

What Yoder reminds us is that only by stopping, only by declining to exercise our will in this way do we force ourselves to look for other possible ways to achieve admittedly desirable ends. Only by declining to use embryos for this research do we awaken our imaginations and force ourselves to seek other sources for stem cells—as may be possible, for example, if recent reports are to be believed, by deriving the cells from bone marrow or from the placenta or umbilical cord in live births. The discipline of saying no to certain proposed means stimulates us to think creatively about other and better possibilities.

One such possibility will, however, be almost as controversial as deriving stem cells from embryos and must therefore be noted. I refer to the possibility of deriving stem cells from germ cells of aborted fetuses. I have opposed the use of embryos for stem cell research, and I also want, in the last analysis, to oppose this method of acquiring the cells, but the reasons are not immediately apparent. On the face of it, after all, this is simply another form of tissue or organ donation from a cadaver. It does not use, or create and then use, a living human being solely for research purposes. Obviously, however, it threatens to suck us back into the situation I described earlier: where every problem becomes, ultimately, the abortion problem. And here, I fear, we cannot so easily separate the issues, although, of course, various procedural safeguards can be put in place to try to assure ourselves that the promised benefits of research do not in any way encourage abortion.

We can clarify our own judgments on the matter by two simple thought experiments that aim to distinguish the several interwoven moral issues. Would we object to research using tissue acquired *only* from spontaneously aborted (miscarried) fetuses? I cannot see why we should, although it is not really very helpful to propose such a source. Would we object to research using tissue acquired *only* from those abortions that, although induced and intended, were ones we thought permissible (however large or small that class might be)? This, at least in my view, is a harder call. But to use for the benefit of others those whom we have already (even if legitimately) condemned to die is so clearly an example of the strong using the weak that I think we should draw back and say no. The life of a human being has been sacrificed in abortion, legitimately by hypothesis, for the good of someone else. As Kathleen Nolan put it, "a moral intuition insists that being used once is enough" (1988, 14). We must challenge ourselves to look for other, better solutions.

The third text is a passage from Stanley Hauerwas, a Methodist theologian: "The church's primary mission is to be a community that keeps alive the language and narrative necessary to form lives in a truthful manner" (1977, 11).

Hauerwas did not mean that Christians are necessarily more truthful than other people. He meant that, when they are doing what they ought to be doing, they worry lest we deceive ourselves, lest we fail to speak the truth about who we are individually and communally, and about what we are doing. This is certainly important for our larger society, and I am quite sincere when I say that in this arena it is an enormous service to speak truly and straightforwardly, to avoid euphemism and equivocation, so that we may together think clearly about who we are and wish to be.

More precisely, I have in mind matters such as the following: that we avoid sophistic distinctions between funding research on embryonic stem cells and funding the procuring of those cells; that we not deceive ourselves by supposing that we will use only "excess" embryos from infertility treatments, having in those treatments created far more embryos than are actually needed[2]; that we speak simply of embryos, not of the "preembryo" or the "preimplantation embryo" (which is really the unimplanted embryo); that, if we forge ahead with embryonic stem cell research, we simply scrap the language of respect or profound respect for those embryos that we create and discard according to our purposes. Such language does not train us to think seriously about the choices we are making, and it is, in any case, not likely to be believed.

I press these three points with some reluctance, because I have the sense that I will be taken to be standing athwart history and yelling "stop!" But it is a risk worth taking. We may easily deceive ourselves about what we do, especially when we do it in a good cause and with a good conscience. We need help if we are to learn to speak truthfully and to face with truthfulness the choices we make, to learn to carry our weakest members, and to seek ethical means to desired ends.

## Notes

1. This chapter is adapted from the author's testimony before National Bioethics Advisory Commission in May 1999.

2. That this is not simply my private suspicion can be seen from the following passage from (Andrews 1999, B5): "Moreover, as embryos become valuable to biotech companies as sources of cell lines, doctors may increase the dose of fertility drugs to insure that multiple embryos are created—in effect, to manufacture more 'excess' embryos."

## References

Andrews, L. B. 1999. Legal, ethical, and social concerns in the debate over stem-cell research. *Chronicle of Higher Education* 29: B5.

Barth, K. 1961. *Church Dogmatics*. Edinburgh: T. & T. Clark.

Hauerwas, S. 1977. *Truthfulness and Tragedy*. Notre Dame, MI: University of Notre Dame Press.

Lewis, C. S. 1947. *The Abolition of Man*. New York: Macmillan.

Nolan, K. 1988. Genug ist genug: A fetus is not a kidney. *Hastings Center Report* 18: 14.

O'Donovan, O. 1984. *Begotten or Made?* New York: Clarendon Press.

Yoder, J. H. 1974. What would you do if ...? An exercise in situation ethics. *Journal of Religious Ethics* 2: 91.

# 14

## On the Elusive Nature of Respect

Karen Lebacqz

Can one speak of having respect for early embryos or for embryonic tissue? If so, what would respect require in this context? That this is a contended issue is evident from several chapters in this volume (e.g., Parens, Baylis, Meilaender, Young). We are used to the concept of respect for persons, but is it meaningful to speak of respecting embryonic tissue?

Several bioethics boards affirmed that the embryo can be deserving of respect. The Human Embryo Research Panel (HERP 1994) declared that, even though the embryo before development of the primitive streak (around 14 days after fertilization) does not possess crucial qualities necessary to be counted as an individual person, it is nonetheless entitled to respect. An earlier bioethics board similarly declared that the human embryo "is entitled to profound respect" while asserting that this respect does not necessarily encompass the full legal and moral rights attributed to persons (Callahan 1995, 39). The National Bioethics Advisory Commission (NBAC) subsequently also used the language of respect when speaking of embryos (1999). Finally, the Geron Ethics Advisory Board (GEAB) established as its first principle for stem cell research that "The blastocyst must be treated with the respect appropriate to early human embryonic tissue" (1999, 31). This history suggests that early embryos or embryonic tissue can be respected.

Yet the matter is not settled. Daniel Callahan took the HERP to task precisely for saying that the embryo is deserving of respect but that it can also be used in research. No criteria are offered, he charged, for determining how to weigh the value of the embryo against the potential good of research. Without such criteria, speaking of respect for the embryo is an empty exercise (Callahan 1995). If we look under the rhetoric of respect, Callahan suggested, the embryo has in fact been stripped of any value at all. To take away value is not to be respectful, whatever the rhetoric. In his view, one cannot be respectful while allowing a kind of use that would not be allowed for persons. Similarly, Gilbert Meilaender argued before NBAC that use of embryos for stem cell research would be disrespectful (see chapter by Meilaender).

We must ask, then, whether allowing research on an embryo is in some way inherently disrespectful: does it strip the embryo of value? Does respect for an embryo mean that it can never be destroyed, as happens in stem cell research?

I believe that one can indeed speak meaningfully of respecting embryos or embryonic tissue, and that criteria for such respect can be established. Specifically, the tissue must not be treated cavalierly, but as an entity with value. Therefore, moral duties are relevant to its treatment, and moral traces of those duties set limits on actions. To make my case I use a form of casuistic reasoning. beginning with the paradigm of respect for persons, I walk through a series of paradigms to see what each might teach us about respect and its possible application to the early embryo (Jonsen and Toulmin 1988). As we move away from respect for persons, agreement about the meaning of respect and its appropriate application diminishes; nonetheless, I believe that by drawing not only on philosophical concepts about the meaning of respect but also on practices that might be deemed respectful, we can locate salient features that prove applicable to the context of the early embryo. I speak of respecting both embryos and embryonic tissue, because the creation of human embryonic stem cells involves use of an early embryo (blastocyst) from which particular tissue (inner cell mass) is derived and manipulated.

## Respect for Persons

The National Commission for the Protection of Human Subjects of Biomedical and Behavioral Research made respect for persons one of three cardinal principles in *The Belmont Report* (U.S. Department of Health, Education, and Welfare 1978). Whereas the commission made room for respect for both autonomous and nonautonomous persons, later tradition generally restricted it to respect for autonomy, thus appearing to limit respect not simply to persons but to autonomous persons.[1] This makes it difficult to know whether respect can be applied to entities other than autonomous persons, such as embryos.

The meaning of respect for persons is elucidated in Downie and Telfer's (1970) classic *Respect for Persons*. These authors maintain that respect for persons includes both an attitude and a moral norm. As an attitude, it implies thinking that something is valuable or estimable. Having respect implies that the thing should be cherished. As a moral norm, it means treating a person as an end and not merely as a means or as something useful for my own ends or purposes (Downie and Telfer 1970, 15). It should be noted that having respect involves ways of thinking and feeling as well as ways of acting.

But why should persons be respected? Kant attributed the distinctive quality of persons that made them worthy of respect, or of being treated as ends, to the ability to reason and the rational will (Downie and Telfer 1970). In short, self-determination or autonomy became central. However, for Kant, self-determination was also coupled with ability to govern our conduct by rules, and it is this rule-giving and rule-governed behavior (auto-nomos) that most clearly distinguished those deserving of respect.

Drawing on this Kantian tradition, Downie and Telfer (1970) hold that to have respect for a person is to make that person's ends our own; it here requires a kind of active sympathy, a practical concern for others. But because persons are rule making and rule following, it also implies that we recognize that other people's rules might be valid and might apply to us. Thus, the attitude of respect includes at least these two components: active sympathy and readiness to hear the reasons of others and to consider that their rules might be valid. We are to try to see the world from the other's point of view (Downie and Telfer 1970).

Under such a characterization, could the early embryo be considered a person deserving of respect? Yes, and no. The embryo (or embryonic tissue) can be considered to have value. It can be cherished. It can be treated not simply as a means to someone else's ends. To the extent that respect for persons requires this general attitude of valuing and the rather vague moral norm of not using another simply as a means to our own ends, respect for persons appears to be able to fit the case of the early embryo or of embryonic tissue.

But the application is difficult if we look closely at what it was about persons that led Kant to consider them deserving of respect. Embryos lack self-determination or rational will. They may have the potential to develop reason, but it does not make sense to speak of respecting their reason, for it is not yet developed. Similarly, embryos are not yet able to be rule-governed beings, and hence it makes little sense to speak of having respect for their rules or for their autonomy. In short, if we ascribe personhood to the embryo, we draw on an understanding of personhood not based on qualities that Kant enunciated. When we turn from embryos to embryonic tissue, the situation is even more complex, as tissue removed from the context of the embryo itself lacks even the potential to develop reason or rule-governed behavior.

Given the fact that some characteristics of respect for persons appear to fit easily the context of the early embryo and others do not, it is not surprising that the debate about whether embryos are persons or can be treated with respect has raged long and hard in the United States. The debate assumed that we know what respect requires because we know what its application is in the arena of persons; therefore, the only morally relevant question is whether the embryo is a person in a moral sense. But there is a different way to approach the issue. We may be better served to acknowledge that embryos do not easily fit Kant's criteria for personhood—and certainly that embryonic tissue removed from an embryo does not—and ask whether there are contexts other than personhood in which respect is nonetheless meaningful. If there are, lessons from those contexts might be brought to bear on the case of the early embryo.

## Respect for Nonpersons?

In Jewish and Christian traditions, God's love is understood as being especially directed precisely toward those who are often denied status as persons in their culture: the outcast, the stranger or sojourner who lacks citizenship, the widow who has lost her social position, the orphan who has no social standing, the poor who are otherwise reviled (see the chapter by Meilaender). Respect is owed not simply to persons, but very precisely to those who are always in danger of being cast outside the system of protection that personhood brings. In such an understanding, an embryo need not be a person to be deserving of respect. Indeed, it may be precisely because it is *not* considered a person that its value needs more urgently to be upheld. The requirement for respect is not diminished.

The term "respect" comes from the Latin *re-specere*, "to look back at" or "to look again" (Webster's *New World Dictionary* 1979). To have respect is to take a second look, seeing below the surface to find the hidden value. As Downie and Telfer noted, it connotes showing honor or esteem, consideration or regard. This suggests that we can speak of respecting a wide variety of things beyond persons: the flag, the ecosystem, religious rituals, cultural practices, scientific data, and so on. It is therefore meaningful to speak of respect in contexts in which we do not have Kantian personhood. A review of some of these contexts illuminates requirements of respect that might apply to embryos and embryonic tissue.

### Respect for Sentient Beings

There are several ways in which respect for sentient beings might be understood. A number of philosophers developed ethical understandings of respect for animals, often linking those notions to animal rights (Regan and Singer 1976; Stortz 1991). If respect is restricted to rights (along the model of respect for autonomous persons), the difficulty becomes specifying what constitutes appropriate animal rights. However, it is not necessary to use rights language to see animals as deserving of respect. Downie and Telfer appear to base their argument not on rights but on duties. When it comes to animals, however, the duty to avoid unnecessary suffering arises out of respect for them not as persons but as sentient beings. Sentience is the basis for the development of distinctive aspects of personhood such as self-determination, and thus it may provide the basis of respect for those who are not fully persons (Downie and Telfer 1970).

However, what it means to show respect for sentient beings requires clarification. Current federal guidelines state that appropriate research be done on animals before doing research on human beings; hence, we have built into the very system of scientific research a requirement that animals be used as research subjects. Although I served on the national commission that affirmed such requirements, I remain uneasy and troubled by an approach that permits humans to use other species for our well-being. When I mentioned this dis-ease to colleagues, they rapidly countered with the fact that most of us eat animals. Is doing research on animals any more disrespectful than other uses of animals, such as for food? If it is not disrespectful to kill an animal to eat it, they query, why should it be seen as disrespectful to use animals in research? Is respect compatible with the use and even with intending or bringing about the death of the one respected?

In an interesting reflection on biblical laws, Richard Hiers maintained that the codes of ancient Israel did indeed provide that death and respect could go together. Humans were allowed to kill animals and eat their flesh, but not to ingest their blood, which represented their life (1996–1998, 134). The life had to be returned to the earth from which it came. This pattern suggests at least one model under which sentient beings might be killed and yet respected.

Another pattern may come from Native American or First Nations peoples. Practices of prayer or chanting or asking forgiveness of an animal that is killed for food suggest that there are ways to be respectful even in the act of killing.

Yet another pattern comes from the work of Temple Grandin, a well-known animal husbandry expert. Grandin (1995) believed that her autism allowed her to empathize with animals, to know how they feel and think.[2] She aspired to bring a keen sense of animals' feelings into animal husbandry.[3] Sacks assessed Grandin's attitude as one that comes close to love (Sacks 1995). Yet I believe that we could also speak of her work as evidencing respect in the sense of care or consideration for the sensibilities of the other. Grandin valued animals even though they were slated for slaughter. Based on her work, we can say that respect for sentient creatures such as cattle and hogs would require several things.

First, pain should be minimized. Noting, for example, that kosher practices were designed with humane intent, Grandin was troubled because cattle were sometimes hung upside down by one leg before slaughter. As the animal was hoisted up, the leg often broke, causing intense pain. Grandin designed a chute for kosher slaughter that allowed the animals to stand upright, thus reducing the occurrence of pain.

Second, fear and stress should be minimized (Sachs 1995). Grandin's ramps and chutes curve, so that animals do not see what awaits them at the end of the ramp. Cattle are able to stay in close proximity to each other, following normal patterns of ambulation. The killing itself is instant, so that the animal has no time to experience fear.[4]

These examples suggest that killing per se is not necessarily disrespectful. It is a question of *how* animals are killed. Respect or disrespect lies not alone in what acts are done, but in the attitude accompanying those acts. (Recall that for Downie and Telfer respect is not simply a moral rule, but also an attitude.) Thus, respect for a living creature may be compatible even with anticipating or designing that creature's death. By extension, use of animals in research is not disrespectful per se, even if it involves their death. What matters is whether they are subjected to pain or suffering, to terror, fear, or stress. Implications for research using embryos are obvious: intending or implementing death is not necessarily disrespectful; respect requires attention to minimization of pain and reduction of fear or stress.

## Respect for Plants

In dealing with the early embryo, however, we are dealing with a living being that does not yet have the physical substrate necessary for feeling or emotion. Concerns for minimizing fear or pain are therefore not terribly relevant, even if they are required by respect. Where else might we look to understand what respect requires where there is little or no sentience per se?

For this, I believe that we can turn first to Barbara McClintock's work with plants (Keller 1983). If it seems strange to think of respecting cattle, it may be even stranger to consider the possibility of respecting corn. Nonetheless, that is precisely what I believe McClintock did. She attended to the individual nature of every corn plant, never trying to force them into a mold. She came to know them as individuals. (At one time she spoke of them as her "friends.") She expected the unexpected: she was open to the possibility that they operated out of rules that were not known and understood by humans. So attuned was she to their variations that she became able to see genetic changes by glancing at the plants.[5] When her ability to see failed her, she described how she needed to "work on herself." She withdrew and sat under a tree meditating until she could return to the plants with proper attentiveness. Her conclusions about gene transposition defied scientific conventions of her day, but many were subsequently substantiated.

McClintock's work suggests several possible meanings of respect. First, it means attention to the concrete reality of the other.[6] Rather than imposing preconceived notions of who or what the other should be, respect means trying to perceive the other in itself. Second, respect requires humility, in the sense that we acknowledge that we may need to "work on ourselves" in order to perceive correctly. To be respectful of life requires a carefulness of vision. Where respect for persons requires respect for the rules of the other and willingness to believe that their rules may be more correct than my own, respect for life more generally might require respect for the ways of the other, and willingness to believe that their ways may have something to teach us and our perceptions may need correction.

### Respect for the Ecosystem

A second source of understanding respect when dealing with less than sentient beings is ecological ethics. Here, respect is variously applied to ethical behavior toward the environment, nature, or creation. Aldo Leopold spoke of using the land with love and respect (Leopold 1949). Roman Catholic Drew Christiansen (1991) noted that the papal encyclical *Sollicitudo rei socialis* speaks of respect for creation as the first of three moral guidelines for dealing with the environment. Protestant theologian Charles McCoy (1991) also urged respect for creation.[7] John Rodman (1977) suggested that we should respect the wild; not just sentient creatures, but land, rocks, trees, and rivers. Do any or all of these uses of respect assist our search for the meaning and application of respect to early embryonic tissue?

Each author's use of respect is nuanced differently. Leopold (1949) spoke of belonging to a community with the land, including the earth and all its creatures. In Roman Catholic natural law tradition, respect for nature means taking account of "the nature of each being and of its mutual connection in an ordered system" (Christiansen 1991, 256).

McCoy argued that when we violate any part of the created order, we harm the whole and hence ourselves. Rodman claimed that it is important that there be a realm beyond our manipulation and control.

None of these authors clearly spelled out what respect means. Johnson (1984) took several of them to task for this failure, and held that a notion of respect for life or for creation or for the wild does not hold up to rigorous philosophical analysis. For instance, what we now consider to be nature or the wild is in fact a product of interventions long ago; all life forms and natural objects adapt to new circumstances. Hence, Johnson was skeptical about an ethic of nonintervention or noninterference in nature, and concluded that an ethics of respect for life that attempts to extend respect beyond persons and sentient beings is unduly vague and cannot be substantiated.

Nonetheless, I think that we can draw some insights from theories that extend respect beyond persons or sentient beings. Two fundamental tenets appear to be at stake in the nuances within ecological ethics. First is an affirmation of the *independent value* of other creatures and of the ecosystem itself. There is a fundamental shift from seeing nature as valuable *for us* to seeing it as valuable *in and of itself*. Thus, respect implies valuing the other. This decenters our perspective: value exists not just because we say so or see it; it exists *apart from* our desires or perspectives. Christiansen (1991) suggested that the fundamental insight of deep ecology as a movement is its assertion of "biocentric equality"—the intuition that all things in the biosphere have an equal right to live and blossom and to reach their forms of unfolding and self-realization.

Second is understanding the interconnection and mutual interdependence of all creation, including humans. Leopold spoke of community, McCoy of covenant. The underlying idea is that all are part of a whole, and that damage to any part of that whole damages every part of it, directly or indirectly. This implies a symbiotic relationship, and to disrespect another part of creation is to harm ourselves, whether or not we realize it.

Kant saw duties of respect as largely duties of omission: the duty not to harm, the duty to leave alone, the duty not to mock others or detract from them, and so on (Nell 1975). Ecological ethics clearly extends such duties of omission to include not just other persons but the entire ecosystem itself: it should not be harmed, it should be left alone to develop according to its intrinsic nature, and so on. But as in the case of Grandin with her animals or McClintock with her corn plants, respect seems to take on a wider meaning: it is not simply a matter of leaving alone or not harming but of standing in awe and making every effort to support the flourishing of the system. It is also a matter of decentering human perspective, recognizing that we may not be as wise as we think.

Thus, it may be possible to speak of respect for that which is not a person, not sentient, and not even yet an individual creature, but a part of a vast and all-encompassing system of nature or creation. Respect implies seeing the intrinsic value of the other, a value not dependent on human valuation but on a larger perspective or on the role of that creature in the entire system. The value of the other is honored by seeing its life as intrinsically intertwined with our own lives. What are the implications of these conclusions for respecting early embryos or embryonic tissue?

## Respect and Stem Cell Research

These reflections permit "a fundamental principle of respect for human life" according to the Geron Ethics Advisory Board (GEAB 1999, 33) to apply at every stage, including stages before sentience or Kantian personhood. What respect requires will differ at each stage. What, specifically, does respect for an embryo require? The GEAB stipulated only that the blastocyst be "used with care" in research "that incorporates substantive values such as reduction of human suffering" (GEAB 1999, 33). Pellegrino (1999) calls this stipulation a "fragile form of respect" that makes the embryo's dignity and value conditional on something outside itself. I attempt to offer a more full account of what respect might require when dealing with an embryo.

The discussion of respect and of respectful treatment of animals and other beings suggests that respect in the context of the early embryo or embryonic tissue would require two things.

First, the embryo or tissue must be valued. The GEAB tried to signal this by suggesting that only research incorporating substantive values would be sufficient to weigh against the value of the early embryo. However, this formulation may not be sufficient. To value something is to believe that it has moral worth in itself, apart from usefulness to us (Thielicke 1970).[8] To respect the embryo is to affirm that the value of the embryo or tissue is *not* dependent on its value for us or its usefulness to us. Respect sees a value in itself beyond usefulness.

If something is respected in this way, it can still be used. Even Kant's formulation of respect for persons does not prohibit persons from being used as means to others' ends, but from being used as means *merely*. In that use, the value of the other must be retained. This is sometimes reflected in suggesting that an attitude of respect or awe (NBAC 1999) should accompany our approach to the entity to be used. To approach something with awe or reverence means that we never become hardened to its intrinsic value, its value apart from us. I suggest that the embryo should not be used cavalierly.

An entity is treated cavalierly if it is demolished without any sense of violation or loss; if it is treated as only one of many and easily replaceable; if its existence is made the butt of jokes or disrespectful stereotyping. Thus, to require that a blastocyst not be treated cavalierly is to require that it be treated as an entity with incredible value; as something precious that cannot be replaced by any other blastocyst, whose existence is to be celebrated, and whose loss is to be grieved.

Second, such an entity can be used in research and can even be killed. To do so is not in itself disrespectful. However, the fact that it can be used and killed does not mean that moral duties no longer hold. W. D. Ross gave us a set of prima facie duties: moral requirements that hold unless there are conflicts among them. These duties include not harming, doing good, being fair, keeping promises, being grateful, making reparations, and improving oneself (Ross 1988, 21). To these one might add the duty to liberate the oppressed or to tell the truth. Prima facie duties sometimes conflict; to avoid harm, we may have to break a promise. Ross gave no hard and fast rules for determining which duties take precedence in such cases. Thus, it may be permissible to do good even at the cost of doing harm.

However, prima facie duties leave what James Childress (1980) called "traces" or "residual effects." Even when we decide that it is permissible to override a duty, that duty still sets limits on our behavior. For example, it may be permissible to do harm in order to do good, but the duty not to harm still holds force: harms must be minimized. Similarly, sometimes injustices must be done; however, the duty to be just still holds, and injustices must be minimized.

In the context of research with early embryos, I interpret the residual effects of prima facie duties along the following lines. It is permissible to use the early embryo in research; however, harm should be minimized.[9] If it is possible to use the embryo without destroying it, that should be the goal. (For example, if cells from the inner cell mass could be taken without destroying the embryo, that should be the practice.) If it is possible to do good for the embryo by giving it continuing life rather than destroying it, giving life should take priority. (For example, if a woman wishes to carry the embryo to term, her desires take precedence over any research goals, no matter how worthy, since her action would preserve the life of the embryo.)[10]

In contrast to Callahan, I believe that it is possible to specify a meaning for respect, even profound respect, for the embryo that will become the subject of research. The fact that an embryo will be used in research does not mean that it is automatically being devalued and disrespected. Just as ancient Israelites exhibited respect by eating animal flesh but not bood, or Native Americans showed respect by first asking the animal's permission and the blessing of the Spirits before killing the animal, so it is possible to approach embryonic tissue with respect that upholds the value of that tissue and sets moral limits on its use.

Whether such respect happens in practice is another matter. In an increasingly secularized society, rarely do we undertake our daily routines in prayerful, respectful, grateful mode. Yet such practices are not impossible. Researchers show respect toward autonomous persons by engaging in careful practices of informed consent. They show respect toward sentient beings by limiting pain and fear. They can show respect toward early embryonic tissue by engaging in careful practices of research ethics that involve weighing the necessity of using *this* tissue, limiting the way it is to be handled and even spoken about, and honoring its potential to become a human person by choosing life over death where possible.

## Notes

1. In my view, this is an unfortunate development (Lebacqz 1999).

2. Grandin (1995) claimed that because she thinks in pictures rather than with language, she can understand how cattle think. She also used connections she saw between autism and animal behavior.

3. Grandin (1995) reported that one-third of all cattle and hogs in the United States are handled in facilities that she designed.

4. In the arena of respect for persons, it is akin to suggesting that prisoners condemned to die, who have lost their civil rights, still should be respected and not subjected to research against their consent, or to research that is unduly painful or stressful.

5. One of McClintock's difficulties communicating with her colleagues may have been that genetic alterations were so obvious to her trained eye that she could not understand why they were not immediately obvious to others.

6. For a philosophical grounding of this concept, see Benhabib (1992).

7. The actual phrase does not appear in the article, but in the synopsis of it, and may have been written by the editors rather than by McCoy himself.

8. Thielicke (1970) suggested that even the most miserable of humans had infinite worth based on "alien dignity" given by God.

9. Since the early embryo is not sentient, the issue of pain does not arise. In use of fetuses, however, as with other sentient beings, pain should be minimized

10. Notice that I do not say here that her desire takes precedence because the embryo "belongs" to her. My argument is not based on ownership, but on the moral duties to do good and to avoid harm.

## References

Benhabib, S. 1992. *Situating the Self: Gender, Community, and Postmoderism in Contemporary Ethics*. New York: Routledge.

Callahan, D. 1995. The puzzle of profound respect. *Hastings Center Report* 25(1): 39–40.

Childress, J. F. 1980. Just-war criteria. In: Shannon, T. A., ed. *War or Peace: The Search for New Answers*. Maryknoll, NY: Orbis Books.

Christiansen, D. 1991. Moral theology, ecology, justice, and development. In: Robb, C. S. and Casebolt, C. J., eds. *Covenant for a New Creation: Ethics, Religion, and Public Policy*. Maryknoll, NY: Orbis Books.

Downie, R. S. and Telfer, E. 1970. *Respect for Persons*. New York: Schocken Books.

Geron Ethics Advisory Board. 1999. Research with human embryonic stem cells: Ethical considerations. *Hastings Center Report* 29(2): 31–36.

Grandin, T. 1995. *Thinking in Pictures: And Other Reports From My Life with Autism.* New York: Random House.

Hiers, R. H. 1996–1998. Reverence for Life and Environmental Ethics in Biblical Law and Covenant. *Journal of Law and Religion* XIII(1): 134.

Human Embryo Research Panel. 1994. *Report.* Washington, D.C.: National Institutes of Health.

Johnson, E. 1984. Treating the dirt: Environmental ethics and moral theory. In: Regan, T., ed., *Earthbound: New Introductory Essays in Environmental Ethics.* New York: Random House.

Jonsen, A. P. and Toulmin, S. 1988. *The Abuse of Casuistry: A History of Moral Reasoning.* Berkeley: Univeristy of California Press.

Keller, E. F. 1983. *A Feeling for the Organism: The Life and Work of Barbara McClintock.* New York: Freeman.

Lebacqz, K. 1999. Twenty years older but are we wiser? *A paper presented to the conference on "Belmont Revisited,"* April 16–18, Charlottesvillc, N.C.

Leopold, A. 1949. *A Sand County Almanac.* New York: Oxford University Press.

McCoy, C. S. 1991. Creation and covenant. In: Robb, C. S. and Casebolt, C. J., eds. *Covenant for a New Creation: Ethics, Religion, and Public Policy.* Maryknoll, NY: Orbis Books.

National Bioethics Advisory Commission. 1999. *Ethical Issues in Human Stem Cell Research.* Rockville, MD: National Institutes of Health.

Nell, O. 1975. *Acting on Principle: An Essay on Kantian Ethics.* New York: Columbia University Press.

Pellegrino, E. D. 1999. Testimony. In *NBAC, Ethical Issues in Human Stem Cell Research.* Vol. III. *Religious Perspectives.* Rockville, MD: National Bioethics Advisory Commission.

Regan, T. and Singer, P. 1976. *Animal Rights and Human Obligations.* Englewood Cliffs, NJ: Prentice-Hall.

Rodman, J. 1977. The liberation of nature. *Inquiry* 20: 83–131.

Ross, W. D. 1988. *The Right and the Good.* Indianapolis: Hackett.

Sachs, O. 1995. *An Anthropologist on Mars.* New York: Knopf.

Stortz, M. E. 1991. Ethics, conservation, and theology. In: Robb C. S. and Casebolt, C. J., eds. *Covenant for a New Creation: Ethics, Religion, and Public Policy.* Maryknoll, NY: Orbis Books.

Thielicke, H. 1970. The doctor as judge of who shall live and who shall die. In: Vaux, K., ed. *Who Shall Live? Medicine, Technology, Ethics.* Philadelphia: Fortress Press.

U.S. Department of Health, Education, and Welfare. 1978. National Commission for the Protection of Human Subjects of Biomedical and Behavioral Research. *The Belmont Report.* Washington, DC: U.S. Government Printing Office.

*Webster's New World Dictionary of the American Language.* 1979. Guralnik, D. B., ed. New York: Fawcett Popular Library.

# 15

## Ethical Issues: A Secular Perspective

Ernlé W. D. Young

Why should someone trained in theological ethics find it necessary to comment from a secular perspective on the moral standing of human embryonic stem cells and germ cells? The reason is that I am a professor of bioethics in a secular university and took a conscious decision years ago to teach ethics rather than morality. The distinction I draw between the two terms is important, all the more because of the way in which they are used synonymously. This is not surprising, since they have the same root in Latin (*moralitas*) and Greek (*ethika*).

Morality is the attempt of individuals, or of groups, to live out in daily attitudes and actions their vision of the highest good. Moral systems, typically, are tied to religious traditions. In general, the highest good has been explicitly, although variously, defined by the world's great religions. *Inter alia*, it is enshrined for Jews in the law of Moses and the prophets; for Christians, in the Hebrew Scriptures and the teachings of Jesus; for Muslims, in the message given Muhammed in the Qu'ran. Within these traditions and the moral systems they have generated, justifications for moral assertions derive from these respective authorities. But these justifications are in what I regard as a "private" language. Only those who are adherents of the same tradition can be expected to understand and accept as authoritative that tradition's religiously based pronouncements on morality.

Ethics, in contrast, employs a common or public language in justifying assertions about prescribed or proscribed attitudes and actions. In the public arena (such as a legislative assembly or a classroom in a secular university), where people may identify themselves with very different religious traditions or with none at all, to justify moral assertions by appeal to the law of Moses, or the teachings of Jesus, or the message of the Qu'ran would be both inappropriate and unpersuasive. The only way to make moral arguments in such a public forum is by using the more neutral language of reason,[1] and by appeal to shared societal values (such as those authoritatively enshrined in the principles of the United States Constitution). Even one's personal moral convictions have to be translated into the common language of ethics in a secular setting. This is simply a mark of respect for those whose religious (and moral) convictions may be different from one's own (Parfit 1984).

A further distinction is that ethics is more at home than morality with uncertainty and ambiguity. Moral systems tend to see things in terms of right and wrong, black and white. Because ethics is invoked precisely when there are disagreements or uncertainties about the rightness or wrongness of proposed courses of action, it is more comfortable with shades of gray. In a less than perfect world our most common ethical choices, of necessity, must be between greater or lesser goods and evils, rather than between absolute rights and wrongs.

It is in this sense that I offer a secular ethical perspective on the issues with which this volume wrestles: What is the moral status of the human blastocyst or embryo? What is the moral standing of the "immortal" cell lines derived either from it or from the aborted ten-week human fetus (Pedersen 1999)? These are serious questions. If an entity has moral standing, it is owed consideration. It cannot be treated any old how, as if it did not have at least some claim on us, however minimal. If a blastocyst or embryo or a cell line derived from an aborted ten-week fetus has moral status, it is owed something. Precisely what it is owed will depend on what sort of moral standing we attribute to it.

Before turning to these questions, it is important to make one point. The decision to use an aborted fetus as a source for potentially immortal germ cells is relatively unproblematic ethically. Only two provisos have to be taken into account. One is that the abortion is legal. The other is that the request to use the abortus to garner stem cells for in vitro culture be strictly separated from the woman's decision to have the abortion. Her decision to allow the aborted fetus to be used for scientific purposes becomes analogous to that of the next of kin to donate organs or tissues for transplantation. Nevertheless, the morality of abortion per se is debatable, hinging as it does on the moral standing of the fetus. The following argument is pertinent to that debate.

My own thinking about the ascription of moral standing was clarified and confirmed by philosopher Mary Anne Warren (1997). Intellectually, I am considerably in her debt. In *Moral Status*, Warren begins by critically examining three unicriterial theories (her term) of moral status that focus on single intrinsic properties—life, sentience, and personhood, respectively—each of which has been claimed to be "the single necessary and sufficient condition for the possession of moral status." She holds that whereas each of these properties may be "sufficient for a particular type of moral standing, treating it as the sole criterion" is implausible and unacceptable. She examines and comes to the same conclusion about two theories of moral status that are based on relational rather than intrinsic qualities: a theory of social or biotic community, and the relational view that the moral status of living things depends on our emotional attachment to them. In the second part of the book, Warren develops what she calls a multicriterial theory of moral status into which she incorporates elements of each of the unicriterial views she found wanting when standing on its own. What follows is a summary of her criticisms and her own position.

Both Albert Schweitzer's reverence for life ethic (which defines life in terms of some special spiritual entity or power) and Paul Taylor's (1986) approach (which identifies living things in terms of their teleologic organization) are intent on extending our moral obligations beyond human beings to the rest of the cosmos. Schweitzer's radical view is that all living things have full and equal moral status. This, as Warren points out, makes brushing one's teeth as problematic as killing flies, cockroaches, and mice, or even members of our own species. Taylor's more moderate view seems to include the principle of self-defense, allowing moral agents to protect themselves against threatening organisms by destroying them. Schweitzer's view ultimately fails because there is no good reason to believe that all living things have a will to live. The argument that all living things have moral status because of their internal teleologic organization is not entirely persuasive either. As long as the moral status of an entity is based entirely on its intrinsic properties (rather than on, say, its place within a biosystem) it is "difficult to demonstrate that life is a sufficient condition for even a modest moral status." Reverence for life is a worthy ideal, as long "as it is not conjoined with the unreasonable demand that we respect all life forms equally. We are not obliged to treat pathogenic microbes as our moral equals" (Warren 1997 49). The more moderate view merely requires that no living organism be harmed without good reason. Admittedly, the terms "pathogenic" and "harmed without good reason" are anthropocentric. Yet it is difficult, if not impossible, to imagine how we can free ourselves of bias in favor of our own species.

Sentience seems like a plausible criterion of moral status, because we assume that it is wrong needlessly to inflict pain or harm on beings capable of experiencing distress or pleasure. Peter Singer maintains that "the comparable interests of all sentient beings be given equal weight in our moral deliberations" (Singer 1979, 19). It follows for him that "all and only sentient beings have interests." For humans to disregard these interests, in the case of animals, is to fall prey to what Singer calls "speciesism"—the moral equivalent of racism or sexism.

But as Warren suggests, this view has serious weaknesses. It is too narrow, excluding from direct moral consideration all nonsentient organisms, as well as species and ecosystems. It explicates the moral status of all sentient beings, including persons, solely in terms of the utilitarian calculus, precluding strong moral rights for individuals. It is inconsistent with such practical necessities as growing food, except through the dubious claim that the lives and happiness of sentient beings that are not self-aware (such as the insects that prey on crops) matter very little to them and may therefore be destroyed. The "sentience plus" view (that sentience is a *sufficient* but not *necessary* condition for having some sort of moral standing) avoids the major objections to Singer's position. "While it offers no account of the relevance of social or ecosystemic relationships to moral status, it erects no obstacle to the environmentalist, Humean/feminist, or other approaches to the construction of such an account. It leaves room for an understanding of moral rights that provides sentient human beings with stronger protections than can be derived from the utilitarian principle of equal consideration" (Warren 1997, 89).

The third criterion for the attribution of moral standing examined by Warren is personhood as set forth by Kant and Regan. The primary advantage of Immanuel Kant's deontologic view that persons (rational beings) ought always and only to be treated as ends over Singer's utilitarianism is that Kant provides individual persons with stronger moral rights. A demand for a categorical respect for the moral rights of individuals is truer to the convictions that are generally held than one permitting those rights to be subordinated to the goal of maximizing utility.

Nevertheless, Warren exposes Kant's theory as vulnerable to a number of objections. First, moral agency (rationality) is not plausibly construed as a necessary condition for any moral consideration, since mere sentience (without rationality) is a sound basis for the attribution of some moral status. Second, there are also grounds for rejecting the view that moral agency is a necessary condition for full moral status. If we take literally Kant's claim that only rational beings are ends in themselves, it would seem that human beings who are not moral agents (because they lack rationality) do not have moral rights. If personhood requires actual moral agency, many sentient human beings are excluded from moral consideration.

Regan (1983) defended a version of this view that at least partially avoids this objection. He believed that most sentient human beings—including some who are not even potentially capable of rational moral agency—have full moral status, as do many nonhuman animals. All beings that are subjects-of-a-life, whether or not they are human, have moral rights, and all of them have the same basic moral rights. "Subjects-of-a-life are beings that possess certain mental and behavioural capacities, in addition to the capacity for conscious experience" (Warren 1997, 107). Being a subject thus plays much the same role in Regan's theory as rational moral agency plays in Kant's.

But Warren brings forward four objections to Regan's view. First, there is the phenomenon of natural predation. Environmentalists recognize that predation is an essential part of every terrestrial ecosystem and therefore cannot be regarded as an evil to be eliminated whenever possible. Even if humans no longer preyed on other creatures thought to be subjects-of-a-life, for us to interfere with natural predation on the same grounds, namely, that animals being preyed on by other animals are subjects-of-a-life, could cause widespread ecological disruption. Second, species and ecosystems that are not subjects-of-a-life can have inherent value. If it is wrong for us to demolish the few remnants of the original Hawaiian ecosystems because of their inherent value, it is wrong for us to permit feral pigs (introduced by humans) to devastate them either, and there may be no alternative to killing them even if the pigs are thought to be subjects-of-a-life. Third is a pragmatic objection: there are insuperable practical obstacles always to treating mice, and rats, and cockroaches as our moral equals. Fourth, there is a line-drawing problem. Just as there is no specifically objective way to distinguish between human infants who have become subjects and those who have not, so, similarly, there is no scientific way to sort sentient animals into those that have (enough of) the mental capacities in question and those that do not. Mosquitoes and monkeys, presumably, have different levels of sentience. But drawing the line between the acceptable eradication of mosquitoes that spread malaria and the unacceptable extinction of species of higher primates becomes a value judgment; it cannot be done on objective grounds. Being a subject-of-a-life is thus neither a necessary nor a sufficient condition for full moral status. We are free to extend full moral status to some beings that do not have all of the capacities that would constitute them as subjects. Conversely, we are free to deny full moral status to some animals that are probably subjects-of-a-life.

The biosocial theory propounded by J. Baird Callicott yields valuable insights (Warren 1997). It permits us to recognize moral obligations to plants and animals and plant and animal species and populations, as well as to such inanimate elements of the natural world as rivers, seas, mountains, and marshes. It encourages us to ascribe equal moral standing to infants and young children who are not yet moral agents, and mentally disabled persons who may never become or never again be moral agents. And it is a more practical theory than those of Singer and Regan. But although the biosocial theory uses a plurality of social and biologic relationships as criteria of moral status, it is nevertheless unicriterial in that it permits only such relationships to serve as a basis for the ascription of moral standing. The biosocial theory provides no satisfactory principle for the resolution of conflicts between different prima facie moral obligations, either those arising from within a single moral community, or those generated by the different moral communities to which one person may belong. Moreover, it allows denial of moral consideration to persons and other sentient beings that are not co-members of our social or biologic communities—a ghost from the Nazi era that continues to haunt us.

Nel Noddings held that moral status is a function of the emotional relationship she called caring (Warren 1997). On this account, it is not necessary for a sentient being already to be part of any of our communities for us to have moral obligations toward it; it need only be *possible* for us to care for it, and for it to respond appropriately. However, her theory has problems of its own. By making moral obligations contingent on the agent's possession of specific empathic capacities, it appears to excuse persons who lack such capacities from moral obligations. Moreover, rejection of moral rules and principles leaves us without guidance in cases where our empathic capacities fail us or have no opportunity to come into play.

Warren concluded that both intrinsic and relational properties play important roles in shaping our legitimate attributions of moral status. But more is called for. This led her to propose her multicriterial view of moral status: one that ties moral standing both to the intrinsic properties of life, sentience, and moral agency, and to important social and ecologic relationships. Accordingly, she propounded seven interactive principles to be used as complementary criteria of moral status (1997, pp. 148–177):

1. The respect for life principle. Living organisms are not to be killed or otherwise harmed without good reasons that do not violate principles 2–7.

2. The anticruelty principle. Sentient beings are not to be killed or subjected to pain or suffering unless there is no other feasible way of furthering goals that are (a) consistent with principles 3–7; and (b) important to human beings or other entities that have a stronger moral status than could be based on sentience alone.

3. The agent's rights principle. Moral agents have full and equal basic moral rights, including the rights to life and liberty.

4. The human rights principle. Within the limits of their own capacities and of principle 3, human beings who are capable of sentience but not of moral agency have the same moral rights as do moral agents.

5. The ecologic principle. Living things that are not moral agents, but that are important to the ecosystems of which they are part, have, within the limits of principles 1–4, a stronger moral status than could be based on their intrinsic properties alone; ecologically important entities that are not themselves alive, such as species and habitats, may legitimately be accorded a stronger moral status than their intrinsic properties would indicate.

6. The interspecific principle. Within the limits of principles 1–5, non-human members of mixed social communities have a stronger moral status than could be based on their intrinsic properties alone.

7. The transitivity of respect principle. Within the limits of principles 1–6, and to the extent that is feasible and morally permissible, moral agents should respect one another's attributions of moral status.

Warren employed these principles to good effect in analyzing arguments for and against euthanasia, abortion, and animal rights. For our purposes, her comments on abortion and human rights (chapter 9) alone are pertinent.

Most abortions take place in the first ten weeks of pregnancy. At this stage fetuses are obviously not moral agents and thus are not accorded full moral standing by the agent's rights principle. Nor are they capable of sentience. Therefore, neither the anticruelty nor the human rights principle applies. However, fetuses are alive and thus have some moral status based on the respect for life principle. Since they are regarded by some people as having full moral status, they may be afforded consideration because of the transitivity of respect principle; but this would be limited by the moral rights that women enjoy under the agent's rights principle. The question is whether we have any independent reason to accord full moral status to zygotes, embryos, and fetuses.

Some hold that the human rights principle ought to apply not just to sentient human beings, but also to the conceptus from fertilization on. To this, Warren responded:

One problem with this argument that the newly fertilized ovum is a human being with equal rights *because it is alive and biologically human,* is that the ovum does not begin to be alive and biologically human only when it is fertilized. Human ova are initially formed in the ovaries of female foetuses; thus, those that are fertilized have already been alive for a number of years. Nor does the ovum become biologically human only when it is fertilized; it has been a biologically human cell throughout its previous years of life. True, it is a haploid cell, containing in its nucleus twenty-four chromosomes, rather than the forty-eight that most (diploid) human cells possess; but this is entirely normal for a human gamete (a sperm or ovum), and does not call into question its biological species.

Few would support the view that each sperm and ovum should be accorded full human rights.

Perhaps the concept is that what begins with fertilization is not biologically human life but the life of a specific human individual. But this claim can also be disputed on empirical grounds. It is not clear that the zygote is the same organism or proto-organism as the embryo that may later develop from it. During the first few days of its existence the conceptus subdivides into a set of virtually identical cells, each of which is totipotent, capable of giving rise to an embryo. Spontaneous division of the conceptus during this period can lead to the birth of genetically identical twins or triplets. Moreover, two originally distinct zygotes sometimes merge, giving rise to a single and otherwise normal embryo. These facts lead some bioethicists to conclude that no individuated human organism exists before about fourteen days after fertilization, when the primitive streak that will become the spinal cord of the embryo begins to form (Warren 1997, 203–204).

Therefore, the respect for life principle accords only a modest moral standing to living things that have no other claim to moral consideration. Women's rights under the agent's rights principle must be permitted to override the slight protection provided to the presentient fetus on this basis.

Nor does much about the moral standing of first-trimester fetuses follow from their biologic humanity or from their status as individuated organisms. Membership of the human species matters for ascribing moral status to an individual who is already sentient, or who was once sentient and may some day return to sentience. However, before the initial occurrence of conscious experience, no being suffers and enjoys, and thus has needs and interests of its own.

We do not yet have enough information to predict with accuracy when the capacity for sentience emerges. There has been a long-standing moratorium on federally sponsored research with human embryos. Yet it seems fairly certain that first-trimester and early second-trimester fetuses are not sentient, since neither their sense organs nor parts of their central nervous system that are necessary for processing sensory information are sufficiently developed. If the first-trimester human fetus is not sentient, it does not come under the protection either of the anticruelty principle or of the human rights principle.

But why should we not extend the human rights principle to *potentially* sentient zygotes, embryos, and fetuses, as well as to already sentient human beings? And how can we exclude presentient organisms without also excluding older human beings who are asleep, unconscious, or temporarily comatose, whose sentience is also (it is contended) merely potential?

The answer to the second question is straightforward. Human infants, children, and adults who are temporarily unconscious are protected by the human rights principle because they have not necessarily lost the capacity for sentience. They are simply not able to exercise it at present.

The claim that the potential of the presentient fetus to become a human being is strong enough to give it full moral status is subject to a *reductio* argument. The unfertilized ovum also has the potential to develop into a human being. The view that potential human beings should never deliberately be prevented from becoming actual human beings implies that not only is abortion morally wrong, but so is contraception (as is masturbation because of the "waste" of sperm). If this view is correct, it is also illicit avoidably to abstain from heterosexual intercourse during periods of probable fertility.

Many people ascribe to human fetuses, even to those in the earliest stages of development, the capacity to suffer.[2] The transitivity of respect principle requires that this empathic response be respected to the extent that this is feasible and consistent with other sound moral principles. Persons who feel empathy for first-trimester fetuses are entitled not to harm them. However, they cannot reasonably demand that others share their belief that these fetuses are sentient, or insist that others must accept the moral conclusions that this belief entails.

The agent's rights principle is more central to the ethics of abortion. Unlike presentient fetuses, women actually are moral agents, and reproductive freedom is both an essential ingredient in the right of women to liberty and necessary for responsible moral agency. Unlike fetuses, women have made a considerable "investment"[3] in becoming the persons they are. The transitivity of respect principle does not override the basic rights of moral agents. Inclusion of infants within the scope of the human rights principle requires no contraction of women's rights to life and liberty, since an infant's physical separateness usually makes it possible for others to care for it should the mother be unable or unwilling to. In contrast, including presentient fetuses in the scope of the human rights principle severely constricts women's freedom (threatening their liberty and well-being) as well as their ability to exercise moral agency.

To summarize, Warren's principles, applied to preembryonic human tissue (blastocysts and cell lines derived from them and from aborted fetuses), yield the following conclusions.

The respect for life principle requires that preembryonic human tissue be treated respectfully, and that it be used only for a serious moral purpose, such as research and development that will eventually lead to improvement of the human condition. Using the blastocyst to develop immortal cell lines falls under this heading. Precisely what it means to treat preembryonic tissue with respect is an interesting question. For the purposes of this chapter, I take it to mean using the minimum number of blastocysts and aborted fetuses possible, and using them only with consent and for the purpose of medically important research and development the scientific design of which is sound.

The anticruelty principle does not apply (the blastocyst is not sentient). The agent's rights principle does not apply (the blastocyst is not rational). The human rights principle does not apply (for both of the above reasons). The ecologic principle does not apply (blastocysts are not necessary to an ecosystem or habitat). The interspecific principle merely reinforces the respect for life principle in this context.

The transitivity of respect principle requires that we show respect for those who are fundamentally and unalterably opposed to the use of *all* human tissue for research purposes and do not dismiss their objections out of hand. However, it also might require those who are so opposed to respect the view I am defending: that the independent moral standing of preembryonic human tissue is so slight and the research being done is so important, that it should be allowed to proceed both with respect and in accordance with the principles of informed consent (the agent's rights principle, as applied to the donors of such tissue).

I am not at all confident that there will be any evidence of mutual tolerance for opposing points of view on this sensitive issue, simply because of the difference between ethics and morality. Despite her extensive use of the term "moral status," Warren's is an ethical position, as are the conclusions I have drawn from it. It depends heavily on reason. The arguments of those opposed to research with human embryos, and to deriving germ cells from aborted fetuses or stem cells from surplus embryos, are moral. That is, they are based on deeply held beliefs, values, and convictions, leaving little or no room for compromise. An inherent tension exists between faith and reason. Unfortunately, it is likely to continue to be expressed here as it has been so regrettably with respect to the related issues of abortion and euthanasia.

## Notes

1. Roman Catholics hold that the moral teaching of the *Magisterium* is fully accessible to reason. Empirically, however, this claim can be questioned. Many reasonable people, for example, do not accept the Roman Catholic Church's teaching that sexuality and the intent to procreate ought never to be separated.

2. This is what Bernard Nathanson did so effectively in the pseudodocumentary film *The Silent Scream*.

3. This is Ronald Dworkin's term (1993, 96).

## References

Dworkin, R. 1993. *Life's Dominion: An Argument About Abortion, Euthanasia, and Individual Freedom.* New York: Knopf.

Parfit, P. 1984. *Reasons and Persons.* Oxford: Clarendon Press.

Pedersen, R. A. 1999. Embryonic stem cells for medicine. *Scientific American* 280(4): 68–73.

Regan, T. 1983. *The Case for Animal Rights.* Berkeley and Los Angeles: University of California Press.

Singer, P. 1979. *Practical Ethics.* Cambridge: Cambridge University Press.

Taylor, P. 1986. *Respect for Nature: A Theory of Environmental Ethics.* Princeton, N.J.: Princeton University Press.

Warren, M. A. 1997. *Moral Status: Obligations to Persons and Other Living Things.* Oxford: Clarendon Press.

# IV

Public Discourse, Oversight, and the Role of Research in Society

# 16

# From the Micro to the Macro

Thomas A. Shannon

As with many developments in bioethics in the past decade, stem cell research and the promises implicit in it hit both the scientific community and the media with a cosmic explosion, leaving in its dust many people—scientists and ethicists, among others—who were both astonished and confused about this latest development. What I want to highlight in this chapter is that this development has both micro and macro issues, and whereas many have focused on the micro issues, such as the status of the organism from which stem cells are obtained, macro issues such as commitment to high-tech medicine and therapies that are directed to the privileged are often neglected. I maintain that although ethical justifications are possible for obtaining stem cells from human embryonic tissue, nonetheless the larger social issues call for at least caution in pursuing this research.

**The Central Micro Ethical Issue**

The most critical micro ethical issue in stem cell research is the source of the cells. These have been obtained in two different ways: from germ cells from aborted fetuses and from cells of embryos not used in in vitro fertilization (IVF).

In the former case, the particular ethical issue is cooperation in the evil of abortion, assuming of course that abortion is a moral wrong. If abortion is not a moral wrong, the ethical problem is ensuring separation of the consent to use the tissue for this purpose from the consent to the abortion, to ensure that the abortion is not coerced. If one thinks that abortion is wrong, one could still contend that researchers are at a sufficient moral distance from the procedure to be able to use the tissue.

Using embryos from IVF clinics or generating embryos to obtain their stem cells presents, for many, more difficult ethical problems. If one's position is that personhood begins with fertilization, one would hold that no human embryo could be used in this way. I wish to develop the other possibility: namely, that although bearing a unique—at least thus far—genetic code and although assuredly human, the embryo at this stage is not a person and thus some interventions can be done.

For me this position has three levels. First, although genetically unique, cells at zygote and blastomere stages are totipotent or pluripotent. That is, they are not yet differentiated or committed to the particular cells they will become in the body—heart cells, liver cells, and so on. Totipotent cells have the capacity to become an entire organism and pluripotent cells various body parts but not an entire organism; hence their obvious desirability for stem cell research. However, the very structure of these cells, while conferring some biologic unity on the developing organism, also strongly suggests absence of a more critical ontologic level of organization.

Second, the developing organism is not yet an individual. Whereas it has biologic unity and organization, its cells can still be separated through twinning or divided through embryo division, and thus different whole organisms obtained. The blastomere can be divided into parts, each of which can become another organism. That is, it is divisible and its parts can become wholes. Such an organism is by definition not an individual. An individual is literally indivisible, or if divided, is divided in such a way that what remains are parts only. The individual is no longer there.

Third, although the National Bioethics Advisory Commission describes this organism as a human life form, I think a more suitable, although perhaps more complicated, way of thinking of the early embryo before differentiation is as a biologic expression of human nature. I base this view on the philosophy of individuation of the medieval philosopher Duns Scotus. Elements in his theory of individuation lend themselves in a particularly helpful way to evaluate morally the status of the blastomere.

The term Scotus uses is "common nature" and it is a part of his larger theory of individuation. Common nature is essentially the basis for the definition of an entity—what all members of a particular class share in common. But what is important for Scotus is that this common nature is indifferent either to being a particular individual or referring to all members of that particular class. Thus, it requires something else—an individualizing principle—to make it a particular being of this class. In addition, this common nature has a unity, but less than a numerical unity. That is, common nature is not an individual being, which would by definition give it a numerical unity, but rather has a unity characteristic of, or common to, members of the entities it defines. Individuation constricts, as Scotus says, the form of this common nature into an individual, rendering this being individual and distinct from all others of the same species. Individuation also renders it indivisible, thus giving it a numerical unity and making it incapable of being divided into two wholes.

We can think of the blastomere as the biologic equivalent of Scotus's concept of common nature, because whereas this entity is genetically distinct from its parents, it is not yet individuated. This does not occur until after the process of restriction is completed, some two weeks after fertilization. To my mind this process is an interesting biologic complement to Scotus's concept of common nature being constricted into an individual. After this process is completed, the cells become committed to being specific cells in specific body parts. This is the biologic beginning of true individuality and marks a critical ethical line (Shannon 1994).

Until the line of individuation is crossed biologically, these cells are indifferent to becoming specific cells or a particular body by virtue of their totipotency; they are not morally privileged by virtue of individuality or, a fortiori, by personhood. They are morally privileged by being human cells, cells that manifest the human genome, and an entity that represents the essence of human nature. Essentially such research would use cells that in fact are the reality of human nature in its most basic form and meaning. Such a presentation of human nature in the blastomere is preindividual and prepersonal. And because this is human nature and not individualized human nature (the minimal definition of personhood),

I believe that cells from this entity could be used in research to obtain stem cells. Clearly, consent must be obtained for this research and the blastomeres must be handled with respect. But ultimately, such research is not research on a human person; it is research on human nature and in principle is morally permissible.

Yet a word of caution must be added here. To use such cells in research is to objectify human nature, to make it a means to an end. Whereas it is clear that, all things being equal, it is ethical to do research on humans, and whereas it is clear that humans can donate body parts for research, it is another thing to generate human embryos exclusively for research. I would not argue that ending the life of such an organism at the totipotent stage is murder, there is no subject of such an act, but this means of obtaining stem cells does reduce the embryo to an object. Therefore, we must be exceptionally cautious about such use and perhaps make it the last resort.

There might be a technical fix to this problem. Research is continuing its attempts to isolate stem cells from adult cells, from fetal bone marrow obtained after miscarriage, or from umbilical cord blood. What is preventing the use of such sources is technical in that the capacity to obtain the cells is not perfected, and financial in that such extraction procedures will be costly. Clearly, efforts in this direction should be encouraged because obtaining stem cells from these sources would resolve a core ethical problem.

### The Macro Ethical Issues

The promise of stem cell research is significant and important. It is possible that these cells can be used for drug development, toxicity testing, studying developmental processes, learning about gene control, and developing specific cells for use with bone marrow, nerve cells, heart muscle cells, and pancreatic islet cells. Further promise is captured in the common description of such cells as "immortal." The hope is that these cells can be directed to develop in ways so they can be either grown into specific tissue or organs, or directly injected into the problem area to replace or compensate for the diseased cells there.

Several comments are in order. First, the promise. One of the characteristics of research into genetics is exaggeration and inflation of claims or, as it is called by some, "genehype." One of the most extreme examples was the claim by a senior scientist at the beginning of the genome project that the success of the project would lead to solving the problems of poverty and homelessness. Everyone can appreciate the ridiculous overstatement of this claim, but other claims are not seen as exaggerations or as significant promissory notes. In a recent article, for example, Daniel Perry (2000) of the Alliance for Aging Research suggests that the approximately 128.4 million patients suffering from diseases ranging from cardiovascular disease, to cancer, to spinal cord injuries, to birth defects may be helped by stem cell research. What we have yet to recognize is the immense and substantive gap between discovery and cure. This is not an argument against stem cell research per se. It is a call to recognize inflated claims that are used to justify commitment of money to a process that is highly experimental and untested. The claim is not the reality, but one would not always know that from listening to discussions of various discoveries.

Second, commitment to stem cell research is a commitment to business as usual in the medical community; that is, to high-tech, very expensive rescue medicine. That is the dominant mode of medicine practiced across much of the United States, particularly in wealthier areas. It is where the money is to be made. Pursuing stem cell research continues this practice and continues to draw large sums of money from other possible uses. As with all other research efforts, particularly in the area of genetics, stem cell research offers great promise for the cure of diseases. But its success will be extremely costly, and its product will also be costly because investors will be seeking an adequate return on their investment.

In addition, as E. Richard Gold pointed out, a commitment to secure property values in human body parts such as embryonic tissue commits us implicitly to specific health policies. First, we will seek cures and turn "away from discovering the underlying social and environmental causes of diseases." Second, we would commit ourselves to a health policy "that holds that health status is improved by access to newer and better treatments." Finally, this policy would suggest that "disease ought to be viewed as an individual problem, specifically a problem of the individual's genetic code, instead of as a social problem" (Gold 1996, 37).

The very difficult social question, is this the way to go with research and medicine? Should we continue down the track of high-tech rescue medicine with its emphasis on intervention and cure, or is it time to have a substantive conversation on other models of medical practice and medical intervention? I do not want to pick on stem cell research, but it is a clear example of another promissory note in modern medicine with the payment to be picked up at the cost of other interventions and research into human well-being, as well as the delivery of health care itself.

Third, who will be the beneficiaries of stem cell research? The rhetoric is that all will benefit. In the meantime, the benefit will be reaped by two groups: those who are insured and whose insurance will cover any resulting treatments, and those who can afford to buy it. Because of millions of Americans who are uninsured, underinsured or whose insurance will not cover such experimental protocols, most will not have access to whatever benefits are realized. In addition, depending on how the science goes, researchers may focus on single-cell genetic diseases because these are easier to identify and target. Again, this narrows the field of application considerably. Research on these diseases using material derived from stem cells may provide the possibility of cure for many who might otherwise be without a remedy; nonetheless this is a significant directing of scarce research money to a small population. Thus, even should stem cell-based therapy prove successful, the number of people who stand to benefit from it are a small subset of the whole population and perhaps even a small subset of all those with genetic diseases.

Fourth, and already alluded to, is the cost of such treatment in both experimental and therapeutic stages. This kind of research is time consuming and labor intensive. Although computers and other automated systems aid tremendously, the main part of the work is both theoretical (understanding genetic structures and planning the research) and practical (carrying out the experiments and studying their results). Well-equipped labs with sophisticated equipment in addition to a highly trained staff are the basic entry requirements. As much of the research will be funded by private capital because of current federal difficulties over the use of human embryos, one can be sure investors will want a return. That return will take the form of expensive therapy. Patients whose incomes are not in the upper 5 percent would not be able to pay for such therapy, as they are not able to pay for many other therapies in our current medical system. Even those who have good insurance plans will have difficulties because of the continued restriction on what will be covered and growing reluctance of insurance companies to fund experimental and expensive therapies. Again the number of possible beneficiaries narrows.

In an early paper on justice and the human genome project (HGP), Karen Lebacqz (1998) suggested that one way to achieve justice would be through some form of price controls in any medications or interventions resulting from HGP research since this research is supported in part by public funding. Private capital investments are an important source of funding for the HGP, but significant monies are also derived from public sources such as the National Institutes of Health. My point is not that such funding is wrong or improper; rather, it is to suggest that we have an obligation in justice to acknowledge these public sources of funding, and a relatively easy way would be to follow Lebacqz's suggestion of some form of price controls.

## Conclusion

I believe that the micro ethical debate over the use of early human embryos is not the key factor in resolving the larger stem cell debate. Although I think that a case can be made for the use of such cells, another more critical variable is the consequence of objectification of human nature in this way. Thus whereas I hold that in principle an argument for the use of such cells exists, the consequences of such use might be more problematic than we realize.

However, I think the more important point is what I have identified as the macro issue, the social context in which such cells would be used. Here I propose that minimally we should be very cautious about going down the path of stem cell research. What we have is yet another promissory note from scientists. Let us at least develop more specific understanding of the therapeutic implications through animal research on stem cells. If one opposes such research on animals, I would note a fortiori that one should also oppose it on humans for exactly the same reasons. As well, developing research on stem cells commits us to the same medical model that is already causing such a complex of problems in health care. Business as usual is not going to resolve our health care crisis. Perhaps it is time to apply the brakes in some areas to try to solve some problems in other areas, such as public health. Finally, those who are most ill and most vulnerable will most likely not have access to the benefits of therapies derived from this research, should those benefits in fact materialize. Insurance plans will probably not cover these treatments and those without insurance or who are underinsured will not have the funds to purchase them. Is focusing on developing expensive cures for a narrow range of diseases the most effective use of public money and social resources?

I am not opposed to the HGP or research deriving from it. I am not in principle opposed to stem cell research. What I am suggesting is a moratorium first to develop some experimental results to see what we are developing and second to force us to think through what kinds of health care reforms we need and how this research might fit into that, if at all.

## References

Gold, E. R. 1996. *Body Parts: Property Rights and the Ownership of Human Biological Materials.* Washington, DC: Georgetown University Press.

Lebacqz, K. 1998. Fair shares: Is the genome project just? In: Peters, T., ed. *Genetics: Issues of Social Justice.* Cleveland: Pilgrim Press.

Perry, D. 2000. Patients' voices: The powerful sound in the stem cell debate. *Science* 287(25): 1423.

Shannon, T. A. 1994. Cloning, Uniqueness, and Individuality. *Louvain Studies* 19: 283–306.

# 17

# "Expert Bioethics" as Professional Discourse: The Case of Stem Cells

Paul Root Wolpe and Glenn McGee

Public policy debates are exercises in rhetoric. The first battle is often a struggle over definitions, and the winning side is usually the one most able to capture rhetorical primacy by having its definitions of the situation accepted as the taken-for-granted landscape on which the rest of the game must be staged. Public debates, however, are not played out on neutral turf. Players make alliances, exercise power, make claims of legitimacy through expertise, and struggle to gain the cultural and political authority to have their perspectives written into policy directives and law. Powerful public movements, such as recent grass-roots opposition in Europe to genetically modified foods, show how large-scale public resistance can recast the debate in terms other than those defined by scientific, academic, or commercial experts.

Research using human embryos, parts of embryos, potential embryos, transgenic embryos, blastocysts and fetal tissue containing primordial cells, and cloned embryos or cloned transgenic embryonic cells—collectively referred to as human embryonic stem (hES) cell research—has been controversial since the identification of the human pluripotent embryonic stem cell in 1998. The hES case is an interesting one because the general impression is that there has been open public dialogue on the issue. In reality the debate has been one among elites who largely managed to shepherd the controversy quickly toward foregone principles and conclusions in which many of the involved experts were invested. In addition, using formal bodies such as the National Bioethics Advisory Commission (NBAC) or the American Academy for the Advancement of Science (AAAS) to consider the matter of hES cell research gives the impression of carefully considered public dialogue.

In fact, public debate has been minimal, and the formal bodies that considered the issue are themselves closely connected to the research federations and institutions that are pressing for the acceptability of hES cell research. The debate was conducted predominantly under a rubric that we term *expert* bioethics, in which issues are framed and conceptualized at a high level of academic sophistication and political authority by groups of highly skilled professionals who are deputized to identify and resolve moral conflict. Most of them are invested in the medical-industrial complex that is vested on one side of the debate. Even NBAC was appointed at the leisure of the sitting president who created that body and who, in his letters charging NBAC with its tasks, made clear that he opposed cloning but favored hES cell research. Twice NBAC issued official opinions that mirror the concerns and opinions of the president.

Reports of the hES controversy in major scientific journals show how the terminology and conceptual framing of the debate by experts are narrow and reflect the concerns of a small, professionally invested elite. Although an appropriate public language to frame the debate is not developed, we believe that a broader and potentially volatile public dialogue is inevitable. We therefore advocate a shift away from public investment in expert bioethics and toward expert-assisted grass-roots debate and discussion aimed at developing, first and foremost, an appropriate and honest public language for the discussion.

One reason that hES cell research, unlike cloning, has not been taken up in any large-scale way by the public media is the success its defenders had in defining the technology early in its development in ways that made opposition and public debate more complicated and difficult to mount. The research was likened at different times by its critics to cloning, to research on human embryos, and to abortion. At each stage, the professional debate began not over the ethical valence of the objection but over the appropriateness of definitions or categorizations themselves. Both supporters and opponents understood a fundamental principle of the politics of rhetoric: whoever captured the definition of hES cell research had won half the battle. If the debate could be configured as being over abortion, including the ban on using human embryos in federally funded research, opponents would in some sense have won. If, on the other hand, supporters could distance the rhetoric from such concerns, they would gain the upper hand.

Ultimately, none of the attempts of opponents to cast the debate in terms that questioned or attacked hES cell research were successful. They could not overcome the considerable authority of the supporters—the prestige of those framing the debate, their institutional legitimacy, and, perhaps most important, their greater access to professional journals whose commentaries and interpretations of the issue informed the lay media. The lay media reported the controversy, but it was not given the profile of other controversies such as cloning, nor were underlying ideologic struggles clearly articulated for the public. The debate was engaged forcefully in journals such as *Science* and *Nature*, but these journals were overt and covert participants in the attempt of researchers to wield scientific expertise as a weapon to control definitions.

### Rhetorical Strategies

Early discussion was mostly limited to the relative obscurity of scientific journals. A cursory glance at relevant articles might lead the lay reader to conclude that the investigation of hES cell research is a technical and scientific matter, or at best one intriguing clue in the larger debate about genetic technology. Yet the debate soon became a political issue, culminating with a charge by the president to NBAC to make a policy recommendation. Politicians, major scientific associations and journals, special interest groups, scientists, and bioethicists became involved.

### Cloning

The hES cell research came on the heels of the controversy over cloning, and the early debate was not so much about abortion as about cloning. The question arose as to whether the research might violate state bans on human cloning or the Food and Drug Administration (FDA) memorandum forbidding clinical use of human cloning technology. Virtually all the popular media reports (and some scientific media as well) quickly accepted the analogy between hES cell derivation and cloning, and often refered to the former as a "form of cloning." Politicians who oppose embryo research quickly pointed to a slippery slope between them. Congressman Jay Dickey (R.-Ark.) told *Nature*, "There are no instances in which I feel the ban on federally funded research on human embryos should be lifted. The language of the ban prevents taxpayer funding for bizarre experiments, such as cloning. Eventually, I could see the embryonic stem cell technology going in this direction" (Butler 1998).

Gun shy from the drubbing taken by the scientific community over cloning, the journal *Nature* wanted to make the difference very clear: "... to describe research using human hES under the generic and emotional description of human cloning, as some reporters continue to do, muddies the waters unnecessarily" (editorial in *Nature* 1999). The first argument over who would control how we talk about hES cell research had been engaged; *Nature* wanted to distance these cells from the term "cloning" to insulate the research from the emotional valence of the cloning debate. Yet the most promising hES cell research involves nuclear transfer from somatic cells to enucleated oocytes. This clearly constitutes an important test of the viability of one part of reproductive human cloning, and embryos produced in that way might eventually be gestated as clone offspring. Moreover, many of the ethical issues, under close inspection, have close moral and technologic parallels to those posed by the science or eventual practice of human cloning. In that sense the controversy over hES cell research would benefit from a comparison with the ethical propositions made in such depth and under such close scrutiny throughout the cloning debates. Hence differentiating hES cells from cloned cells was an important move by its supporters to avoid becoming embroiled in the human cloning issue.

## Pluripotence versus Totipotence

The journal *Science* noted the public confusion between cloning and hES cells obtained through nuclear transfer; however, the latter is not cloning because the resultant cells are not necessarily totipotent (Solter and Gearhart 1999). The degree to which they may be totipotent is still uncertain and depends on definitional technicalities. The derivation of hES cells and development of nuclear transfer technology cast doubt over what words such as totipotency and pluripotency really mean, and whether their definitions should be modified in light of new conventions about the power of DNA in cells of various kinds and stages of development. However, *Science* maintained that an hES cell derived from a blastocyst of any kind is not totipotent because it cannot be *directly* implanted into a uterus and grown into a conceptus. It may not be too long before that is no longer the working definition of totipotence.

The importance of the issue of totipotence was underlined by Michael West, founder of ACT, a company that claimed to have produced a hybrid cell where human nuclei were inserted into the ova of cows. The hybrid cells were justified by ACT because, unlike cloned cells, they are not totipotent and thus are more like cells than embryos. West thus concluded that the combination of cow and human was not a moral issue at all since "these cells cannot become human beings" (Alper 1999). He further claimed that when nuclear transfer is used to make the embryonic source of hES cells, the resulting embryo is itself not really a human embryo, because its creation does not involve either conception or, in the case of the cow ova, enough human parts.

The debate about totipotency and pluripotency is carried on by scientists with authority and clarity that suggest that well-understood standards exist for such matters. Many of those working on the research insist that existing definitions of the power and properties of cells are simply a matter of objective observation, and that making reference to these definitions can resolve or at least clarify associated moral issues. But no real clarity is found in the history of scientific study of the embryo, or even in current canonical works on the properties of embryos and cells. Discovery of the hES cell and research into its powers and properties—as well as those of cells that that are created or hybridized using hES cells—just raise more questions. The most plain conclusion to be drawn from debates about totipotency, pluripotency, and the power of hES research to create clones or embryos is that society has begun to make up new definitions for the powers of cells (McGee and Caplan 1999a). The process of deciding who will refine, reform, or reify definitions of these cells is a sociomoral exercise that has implications for the broader battle for or against hES cell research.

## Abortion and Federal Funding

By denying the relationship between hES cell research and cloning, experts managed to avoid emotionally charged cloning rhetoric. However, the more serious charge was that hES cell research required destruction of embryos, and violated the ban on use of embryonic tissue in federally funded scientific investigation. The ban was closely tied to the abortion debate, a debate in which the scientific community was also loath to engage. If hES cells were identified in the public mind as embryos or even as totipotent cells, great pressure would be applied to invoke the ban. The effect of such a measure would be to stifle federal support for hES cell research, which would increase both the burden on researchers to generate quick clinical applications and intellectual property, and the stigma associated with researchers who would be "complicit" in destroying embryos (Robertson 1999). "Embryo research" as a moniker would make it far more difficult for these researchers to enlist the support and funding of patient advocacy groups who might straddle the fence on abortion. [In early August 1999, the American Cancer Society (ACS) withdrew from Patient's Coalition for Urgent Research (Patient's CURe), a coalition of more than thirty patient groups that lobbied Congress to support hES cell research, due to pressure on ACS from officials of the Roman Catholic Church.]

To head off such objections, the scientific community was quick to claim that hES cells are not themselves embryos and so research on them does not violate the law. Bioethicists, including one of the authors (GM) participated in that effort at public hearings held by Patient's CURe in government testimony, and in the professional literature (McGee and Caplan 1999a,b). In January 1999 the U.S. Department of Health and Human Services (DHHS) issued a legal opinion saying that hES cell research does not fall under the federal ban on human embryo research because such cells do not constitute an "organism" as described in the legislation. As long as researchers did not themselves destroy embryos, the cells could be used, because "even if the cells are derived from a human embryo, they are not themselves a human embryo" (Wadman 1999a).

Yet, even if the cells were not embryos, their sources were. The DHHS opinion involves sleight-of-hand to the effect that because the researchers did not *extract* the cells from embryos, they are not considered engaged in embryo research. Frank Young, a commissioner in the Food and Drug Administration under President Reagan, pointed out the strange kind of reasoning the scientific community was employing: "To say, on the one hand, that you cannot support the deliberate destruction of living human embryos to harvest their stem cells, but that you will, on the other hand, pour millions of taxpayer dollars in support of research that you know can only take place using materials derived from that destruction, is an exercise in sophistry, not ethics" (Wadman 1999b). Michael West also made the point that researchers in his group were not destroying embryos when they made and then destroyed cow-human embyro-like organisms, because a cow-human "embryonic combination" is not viable and thus, they claimed, cannot be called an embryo.

Recognizing the difficulty of arguing that research on hES cells derived from destroyed human embryos does not support that destruction, Arthur Caplan hold that using the term "embryo research" to refer to work in which embryos used are donated leftovers is a misnomer. "The embryos existed, were not intended to be used, and were not created for the purpose of research" (Brower 1999, p140). Researchers removed cells from donated embryos and thus Caplan did not regard them as viable embryos. "They were not created for the purpose of research, which would be both forbidden by US law and morally objectionable." In other words, embryos that are leftover, that is, not intended for implantation, are by definition nonviable and therefore are not embryos that would fall under the ban. Many would disagree that human embryos cease to be embryos simply because no one wants to impant them. But, more important, the view that because the embryos were not created for research they can be viewed as cell donors does not necessarily follow. Moreover the fact that donated embryo are destroyed and are nonvoluntarily donated surely complicates donation in the view of those who object to the research on the grounds that embryos are vulnerable human subjects that should not be experimented on.

One proposal that gets around this objection is that embryos, even if they are vulnerable subjects and even if they exist as full moral persons, are not destroyed in the process of hES cell derivation. Because a blastocyst's genetic information is not destroyed in derivation or subsequent use of the cells, recreation of the original donor embryo might be possible using nuclear transfer of recovered, recultivated, donated cells into an enucleated egg (McGee and Caplan 1999b). This was proposed as an illustration of how difficult it is to identify the destruction of embryos in hES cell research, and its authors concede that recreation of a destroyed embryo in no way diminishes the violence involved in transferring DNA from an original embryo into a new cell. Still, the example has been cited widely by experts as evidence that ethics are on the side of mining particular embryos and embryonic tissue, however irrevocably, for hES cells. The definition of "killing" was altered to exclude destruction of potentially reproducible embryonic individuals, eliminating the moral problem in view of research supporters.

## The Politics of hES Cell Research Rhetoric

Guided by interested scientists and ethicists, scientific media steered the debate carefully down the middle road and defined hES cells out of all problematic categories: they were not embryos, they were not cloned cells, they were not totipotent. They were simply cells. The media never managed to get a handle on why research ethics were important, or to mobilize the public into a meaningful discussion of the issue. Despite the work of antiabortion interests in the Roman Catholic Church, who enlisted their most public intellectuals to fight the battle, hES cells never became a significant national issue.

In the meantime, bodies that were charged with making policy or advising policy makers on the issue were composed overwhelmingly of scientific elites and experts drawn from the research community. The president asked NBAC to produce a quick report and policy recommendations. Harold Varmus, then director of the National Institutes of Health (NIH), noted that he did not have to wait for NBAC's report for NIH to make its decision (Marshal 1998a). Given the total lack of legislative or regulatory power allocated to NBAC, and its rubber-stamp report that did not change law or policy about cloning, it was a foregone conclusion that little could come of NBAC's expert recommendations. As noted above, few in Congress or the media doubted the outcome of NBAC's deliberations, given the scope of its task and the context of its appointment. The AAAS also put together a panel to discuss and make recommendations on hES cell research ethics, and issued a report that mirrored, no surprise, the editorial calls of *Science* for the research.

Special interest maneuvering was going on behind the scenes. In response to a plea from the House pro-life caucus, a group of seventy congressmen and seven senators wrote a letter to Donna Shalala, Secretary of DHSS, protesting the department's decision to fund hES cell research. In response, a group of thirty-three Nobel laureates wrote to President Clinton and the U.S. Congress in March of 1999 to urge the government to permit such research.

But the real work on establishing the legitimacy of the research was being done in the federal agencies tied to research interests. The use of elites to give the imprimatur of ethical review on research has become *de rigeur* in science. Wary of the controversy, Varmus proposed that an outside committee of experts review all hES cell-related grant proposals to square them with the criteria by Congress (Marshall 1999). When Michael West was asked why he released preliminary results on his bovine-human cell hybrid to the media without the full scientific data that would prove to skeptics that it was in fact such a fusion, he replied that he wanted to develop the technology, but was worried about the reaction of the public and the applicable law. "So I decided," he remarked, "let's talk about the preliminary results. Let's get NBAC to help clear the air." Despite its lack of legislative or regulatory power, in other words, NBAC came to be seen by many in the scientific community as the de facto ethical deliberative body for controversial scientific research (Marshall 1998b).

Other countries are also using deliberative bodies for a similar pupose. In Germany, it is illegal to use hES cells from spare eggs or embryos, as the Embryo Protection Law confers full human rights from the moment of conception. The law includes cloned embryos as well. However, there is no law against developing pluripotent human embryonic germ cells from aborted fetuses. Germany's basic research funding agency, Deutsche Forschungsgemeinschaft, called for establishment of a central committee to assess the issue, as well as open dialogue at the European level (Abbot 1999).

In Britain, the government receives recommendations on these issues from the Human Genetics Advisory Commission and the Human

Fertilisation and Embryology Authority. These groups sought public comment through one of their public "consultations," increasingly part of British debate about bioethics, and found that whereas 86 percent of people who commented were against human reproductive cloning, only a fraction wanted to limit hES cell research, and the recommendation of the two commissions reflected these positions (Marshall 1998c). Nevertheless, in a bow to anticipated public debate, Britain imposed a temporary moratorium on certain types of hES cell research. It already issues licenses for those wanting to do research on embryos in the first fourteen days of development (a limited number of such experiments are permitted), and the advisory commission recommended a similar strategy of research licenses for cloning and hES cell research. France's National Bioethics Committee recommended loosening the ban on research on human embryos to allow cultivation of hES cells (Butler 1997).

Private industry is using the same tactics to preempt ethical challenges. Worried about public opinion on their hES cell work, Geron engaged its own ethics board to make recommendations. Smithkline Beecham, one of many large pharmaceutical corporations likely to do hES cell research, held special ethics hearings in which it too searched for a way to find acceptable language with which to promote this work.

### Who Should Make Bioethical Decisions?

The public debate about hES cell research reflects an odd shift from the equally problematic debate about cloning. Regarding cloning, experts struggled to keep up with an expanding maelstrom of public fear and expressions of anger. Institutions were asked by their contituents what their position was and what actions they were going to take. In contrast, the political system and its participants at various levels were ready for the hES cell debate, and the last thing they wanted was another Dolly controversy. Public understanding of the phenonemon must be prefigured and the parameters of the debate defined before the science becomes well known and widespread. The lesson gleaned by the experts from the cloning controversy was simple: *in modern biotechnologic controversies, public debate must be shepherded and fostered by an elite that is prepared to seize rhetorical primacy, and to mold existing institutions or create new ones for that purpose.*

The result of the cloning debate on bioethics as a whole (it is not too early to say) has been paradoxical. After cloning, bioethics plays a more important role than ever in the discussion of science; yet its role has been largely assimilated into the political model of Camelot, in which the philosopher is further elevated above the people and the model of discourse further rarified. The result is NBAC, whose decisions are based on the testimony of elites (many of whom have interests in the medical-biotechnical enterprise), whose membership represents overwhelmingly one side of the politicocultural spectrum, and whose publications cannot be understood by more than a fraction of the public.

If the current model of expert bioethics, somewhat amplified by the cloning controversy, is not desirable, what is the proper role of the bioethicist in matters public? What is the task of the philosopher, social scientist, or clinician with skills in moral matters? The hES cells in all their complexity provide, we believe, an obvious opportunity to test a new model for the role of bioethics in public debate. Before this can happen several key claims must be accepted by elites in the field. First, although bioethics as a discipline prides itself on its openness to public participation, few participants have any mastery whatever of the implements of public debate, including rhetoric, media, and politics. Bioethics must grow more familiar with these skills, while avoiding the sale of its soul or similar distortion of purpose. Second, the role of language in public debates about ethics is an important one, and skills of discerning moral and scientific valences of particular words are crucial. The effort to frame hES cell research in clear language that neither prejudices unnecessarily nor misleads is the most important moral effort now under way, and the role of the bioethicist should be as critic of, and not hand-maiden to, expert efforts to smuggle values into the process. Bioethics can be the proctor of public debate only if it plays this explanatory role with more skill of discernment: rather than accepting the debate at face value, we need a bioethics that questions the false or patently political framing put to NBAC or other bodies of discussion and debate. Third and most important, bioethics must be a public activity itself, written more and more with a broad audience in mind. Even where this goal undermines the first and second goals, it should be reflected in a commitment of bioethics mentors to educate a new generation of scholars whose ambitions for writing and thinking extend beyond a small group of philosophical or social scientific peers.

## References

Abbott, A. 1999. Don't try to change embryo research law. *Nature* 398: 275.

Alper, J. 1999. A man in a hurry. *Science* 283: 1434–1435.

Brower, V. 1999. Human ES cells: Can you build a business around them? *Nature Biotechnology* (Feb. 17): 139–142.

Butler, D. 1997. France is urged to loosen ban on embryo research. *Nature* 387: 218.

Butler, D. 1998. Breakthrough stirs US embryo debate. *Nature* 396: 104.

Editorial. 1999. Towards the acceptance of embryo stem-cell therapies. *Nature* 397: 279.

Marshall, E. 1998a. Use of the stem cells still legally murky, but hearing offers hope. *Science* 282: 1962–1963.

Marshall, E. 1998b. Claim of human-cow embryo greeted with skepticism. *Science* 282: 1390–1391.

Marshall, E. 1998c. Britian urged to expand embryo studies. *Science* 282: 2167–2168.

Marshall, E. 1999. NIH plans ethics review of proposals. *Science* 284: 413–415.

McGee, G. and Caplan, A. 1999a. What's in the dish? Ethical issues in stem cell research. *Hastings Center Report* 29(2): 36–38.

McGee, G. and Caplan, A. 1999b. The ethics and politics of small sacrifices in stem cell research. *Kennedy Institute of Ethics Journal* 9(2): 151–165.

Robertson, J. 1999. Ethics and policy in embryonic stem cell research. *Kennedy Institute of Ethics Journal* 9(2): 109–136.

Solter, D. and Gearhart, J. 1999. Putting stem cells to work. *Science* 283: 1468–1470.

Wadman, M. 1999a. Embryonic stem-cell research exempt from ban, NIH is told. *Nature* 397: 185–186.

Wadman, M. 1999b. White House cool on obtaining human embryonic stem cells. *Nature* 400: 301.

# 18

## Stem Cells: Shaping the Future in Public Policy

Margaret R. McLean

A fitting assessment of human stem cell technology mandates that it be seen in light of how it might be used if it meets the bars of safety and effectiveness. Although research involving human embryonic stem (hES) cells will initially improve our understanding of basic human embryonic development and pathology, these cells are harbingers of a revolution in medical therapeutics in which individual replacement cells and tissues will be used to treat myriad degenerative diseases. In addition, because of their ability to undergo prolonged undifferentiated proliferation—so-called immortality—hES cells are potent tools for genetic germ line interventions. The "bigger picture" (Parens 2000) of hES technology, therefore, encompasses both the potential generation of transplantable tissues and, in combination with nuclear and gene transfer technologies, the possibility of reprogenetically shaping children.[1] It is not an exaggeration to say that no corner of medical practice will remain unaffected as medicine shifts its sights from organs and systems to genes and cells.

Stem cell technology's bigger picture presents unprecedented public policy and regulatory challenges. The current genetic revolution and the steady march of biotechnology deeply affect lives, relationships, ideologies, and social structures. How to respond to the challenges of "the biotech century,"[2] including the balance between technologic development in the public and private sectors and extent and type of regulation, is increasingly important to citizens and of increasing concern to governing bodies and regulatory agencies. These policy and regulatory challenges emerge from the very nature of biotechnology and reprogenetic research and development. Unlike earlier big science projects (for example, space exploration and the Manhattan Project), new biotechnologic developments do not rely on large-scale, centralized societal structures and public funding, but on decentralized, punctate, often privately financed systems. Indeed, the entire field of reproductive medicine remains scantly regulated in the United States. The decentralized nature of biotechnology in general and reprogenetic technology in particular allows for great potential benefit—for example, speed—but also an increased opportunity for abuse—for example, unwarranted secrecy. Stem cell technology places control of the biologic processes of aging and disease as well as germ line genetics into the invisible hand of the market and the fleshy hands of individuals, making public policy formation much more complex than for older technologies that required systematic societal involvement.

Recognizing these challenges, and relying on a consideration of justice as "fairness of access" that pays attention to the social lottery,[3] five ethical principles are suggested as a framework for policy formation within a just society.

## The Dolly Effect

As research proceeds, opportunity arises for prospective deliberation of normative and social issues attendant to hES cell research and development. Prudence would dictate that it is best to avoid the Dolly effect; that is, attempting to close the ethical-legal door only after the sheep has left the barn. Dolly's unanticipated debut left the public and the bulk of the scientific community bereft of a framework for considering ethical and policy issues of nuclear transfer technology and cloning. As the Ethics in Genomics Group asserts, "[a]ttention to the direction in which cloning research was headed before Dolly's creation would have better served society than the overreaction which ensued" (Cho et al. 1999, 2087).

The current state of stem cell technology presents an opportunity to avoid the hysterics of the Dolly effect and to engage in broad debate of essential ethical and social issues. Research on stem cells is of such critical importance that responsible citizens should be aware of the current state of the technology and its implications for human biology and health. The public can begin to understand and fully examine what is at stake if endeavors are made to explain the nature and potential medical application of this science and to illuminate key moral, religious, and social concerns. There is a clear need for thoughtful consideration of the impact of biotechnologic innovations including hES cells on the values, commitments, and institutions that nurture both individuals and communities. The fundamental nature of this research imparts a high degree of moral gravity and mandates that ethical evaluation be integral to public policy formation.

It also seems best to avoid a second aspect of the Dolly effect—the rhetoric of inevitability. Often technologic advances are characterized as "inevitable" by both scientists and the public: "[T]he use of reprogenetic technologies is inevitable. It will not be controlled by governments or societies or even the scientists who create it" (Silver 1997, 11). In this view, science is unstoppable and, as such, is not to be the object of ethical concern or stringent regulation. This sense of fatedness profoundly limits our thinking. It is crucial to acknowledge that the possible, however tempting, however frightening, is not the inevitable.

### Science—Public or Private?

Because laws in many countries, including the United States, preclude public funding for human embryo or fetal research, human cell research has steamed ahead in a handful of privately funded labs. The panic-drenched, reactive atmosphere of the Dolly effect raises questions about the wisdom of it remaining confined to private, commercial enterprises. The extraordinary medical potential of hES cells to treat or cure everything from Alzheimer's disease to paralysis imposes the responsibility to consider openly the societal reverberations of the basic research on and medical use of these cells and the proper form of public policy.

To this end, scientists are compelled to include ethical reflection and research integrity in the scientific agenda. But considerations of stem cell technology are too far-reaching to be left to scientists or to professional moral philosophers or theologians alone. Meaningful dialogue about the advent and application of hES cell technology between scientists and the public is essential. "Meaningful dialogue" means conversations that are mutually informative, honest, thoughtful, broadening, and potentially transformative. If such conversations about human stem cells force reexamination of metaphysical questions concerning the nature of human personhood, the extent of human control over life, and humanity's place in the natural world, among other issues, our time is well spent.

Since there was neither public debate nor citizen oversight of initial forays into hES cell research (White 1999; Trauer 1999), the imperative for public deliberation is deeply compelling. It is notable that the Geron Corporation Ethics Advisory Board (GEAB 1999) invited public discourse on the ethical issues emerging from hES cell research. However, research protocol review and ethics review within a private company such as Geron are necessarily private. It is difficult to imagine that all the cards are on the table when lucrative patent rights and hefty shareholder returns are at stake. Meaningful dialogue in the public arena simply cannot be done in a context of proprietary information and profit enhancement. A true debate over public policy demands that both information and deliberation in fact *be public*. It is critical to move hES cell activities and attendant concerns into the public spotlight where they can be broadly deliberated and the research and its application supervised. This is perhaps the strongest argument for government funding, which requires public discourse and access to information generated by hES cell research.

In addition, the for-profit mode of the market necessarily influences research direction and access to products. Huge profits are to be made if the transplantation and reprogenetic dreams for stem cells come true. As Lisa Sowle Cahill (2000) notes:

The individual rights of investigators, investors and companies to sell biomedical tools enjoy a priority in our legal and political system that is unmatched by the right of other members of society to a decent minimum of health care, much less by practical means of structuring behavior patterns so that they contribute to the common good, and further a humane, holistic approach to health, illness, suffering, finitude, scarcity, and social interdependence. (134)

How these profit rights are to be balanced with concerns for individuals, human health, and a just and humane future is a prime challenge for public policy formation.

## Ethics and Public Policy

People of good will disagree about ethical boundaries, private beliefs, and public policies that ought to govern hES cell research, development, and application. That different moral commitments and belief systems cannot be bridged easily is a frequent challenge to policy development in a pluralistic society. What is required is serious civic conversation about points of consonance and dissonance, benefit and burden. The goal is to cultivate policy that adequately mirrors shared visions of human health and flourishing, promotes society's best interest, and vigorously pursues social justice.

In deliberating policy regarding human stem cells, three observations are important (Shapiro 1999). The first is the pervading uncertainty regarding which rival moral points of view ought to shape public policies. A necessary duality to civic conversation involves concern with what ought or ought not be done and concern for how to determine what ought or ought not be done. The second observation is that we cannot (indeed, ought not) escape the tension that characterizes the current situation in which the justifiability of many ethical claims remains dubious to significant segments of the community. Public policy should strive to be the least offensive to the most persons. The final observation is that the set of optimal ethical views—those that produce the most reputable, responsible and redeeming outcomes—are not likely to remain fixed in the fluctuating circumstances of the biotechnologic age. It is inescapable that the scorching pace of biotechnology in general and hES cell research in particular are destined to create new ethical concerns and misgivings and a penchant for societal control.

The spectacular debut of hES cells immediately rekindled the fiery debate concerning the use of human embryos and fetal tissue for research. Investigations that provide no benefit to the embryo or fetus raise serious questions about the relative importance of treating or curing disease and respecting developing human life. Even thicker ethical concerns are raised by the deliberate creation of research embryos solely for investigative purposes.

However, embryonic and fetal sourcing may become less necessary. First, hES cells are regenerative, and existing cell lines may be of sufficient quantity and quality to produce the required cells if random genetic mutations and cell senescence can be avoided. If such proves to be the case, future need for further isolation of cells from embryonic or fetal tissue may be limited. Second, and seemingly more promising, is the spate of reports of success in isolating and channeling adult stem (AS) cells into particular cell types, including murine blood cells and human bone.[4] Recent research in mice suggests that stem cells taken from the adult brain can be coaxed into a wide variety of tissues, including liver, heart, and muscle (Clarke et al. 2000). In humans, liver cells were derived from circulating bone marrow stem cells, giving rise to speculation that it may be possible to repopulate livers damaged by hepatitis, drugs, or alcohol with healthy cells derived from a patient's own marrow stem cells (Theise et al. 2000). The potential of AS to convert into myriad cell types may eliminate the ethical dilemma inherent in obtaining stem cells from embryos.

Even those adamantly opposed to research involving embryonic or fetal stem cells do not deny the unprecedented potential benefit of cell therapy and tissue regeneration. Hence, if the sourcing moves from embryo to adult, it seems reasonable to assume that ethical and policy questions will shift to the scientific and therapeutic potential of this technology. In anticipation of such a sourcing detente, I focus on the question of access to the benefits of stem cell technology.

## Fairness of Access

Public policy often seeks to regulate behavior based on socially desirable outcomes. A speed limit of ten mph intends the prevention of harm to school children by an imprudent driver. Both the National Institutes of Health (1999) and the National Bioethics Advisory Commission (1999) appealed to consequences to maintain that hES cell research should proceed. Nonetheless, careful consideration ought to be given not only to the intended but also, and perhaps primarily, to the unintended consequences, especially with regard to burdens placed. Circumvention of the argument from inevitability is also necessary as citizens consider precautions to ensure that neither going forward nor staying put harms ourselves or others in an effort to heal, protect, and benefit.

Because human stem cell work portends such revolutionary human benefit, we ought to worry that the benefit will be distributed unjustly and further privilege the monied and powerful at the expense of those on the socioeconomic margins.[5] The tattered backdrop of our current booming national economy includes 46 million uninsured Americans who lack consistent access to the basics of health care and 40 million, including 1 in 5 children, living in poverty. Privatization of health care in the form of for-profit health maintenance organizations has widened the gap between the medically well-off and the medically indigent, leading to grossly inadequate care for those without the bases for access; that is, money and transportation. Given our country's growing economic divide and the fact that private companies are riding the leading edge of biotechnology, it seems likely that, left undisturbed, stem cell technology will be available to some but not to all. This portends further stratification of human health and well-being within the richest country in the world.

The problem of just distribution of medical resources is not new, but the unprecedented promise of stem cell technology raises the stakes. Respect for the dignity of the human person imposes a communal obligation to treat disease and maintain individual and societal well-being. Every human being is a person of worth to be treated always as an end and never solely as a means to someone else's or society's ends. Because health is a social good and necessary for human flourishing, policies and procedures ought to seek to make stem cell technology available in ways that are responsive and responsible to persons and to society as a whole. Against those who argue solely for *equality of opportunity* in access to the benefits of medical innovation, let me suggest an ethic centered on *fairness of access* to the conditions and commodities necessary for human health and well-being, including stem cell technology.

Although equality of opportunity is an important value, it focuses only on limitations imposed by legal and formal barriers of discrimination (Buchanan 1995). However, to sustain human health, attention must be paid not only to legal barriers of discrimination but also to the social lottery that leaves some unable to grasp the ring of opportunity. Fairness of access removes the socioeconomic blockade imposed by the social lottery to health care and levels the playing field. Opportunities for equality of well-being can be secured only through access to those opportunities.

Present public commitments ought to be modulated by their effect on those who are the weaker members of society, especially children and those left poor in money, health, and access by the spin of the social lottery wheel. Power ought to be brought to bear to protect and advance the interests of such vulnerable persons by constraining action to what tangibly benefits the marginalized.[6]

## Shaping the Future

Stem cell and nuclear transfer technologies stretch human power so that future circumstances are subject to present discretions, desires, and duties in unprecedented ways. Hence, justice is not only concerned with contemporary resource distribution but also enjoins responsibility for how future generations are to live. Transgenerational justice imposes present self-limitation in the interest of the life and health of future generations. In turn, public policy formulated for the sake of a just future mandates that consequences of our present actions, both public and private, be appropriate to the flourishing of future generations. The seventh-generation rule obtains: we should consider the consequences of what is done today on each of the next seven generations.[7] Moral wisdom comes through having regard for the interdependence of present and future interests.

Success would be the development of public policy consistent with principles of social justice, especially fairness of access, and responsibility for the future. In view of the unprecedented and uncharted scientific and medical benefits that may result from research on human pluripotent stem cells, basic policy components would include the following:

1. Primary public understanding of stem cell and nuclear transfer technologies and the promises and perils of each.

2. Opportunities for vigorous, honest public debate with all the cards on the table.

3. Public funding of research with attendant public review, oversight, and accountability.

4. Guaranteed fairness of present and future access to the benefits of stem cell technologies, with privilege given to vulnerable persons and communities.

5. Development of standards of excellence for stem cell technology that are consistent with the full scope and goals of health care, just access and future sustainability.

## Conclusion

White-knuckled, we are crossing medical frontiers at break-neck speed. Stem cell technology holds the promise not only of increasing human health and life spans but also of changing power structures and fundamental notions of human personhood, moral status, and mortality. It is important that we do not prematurely or unwittingly slam the door on scientific advances that can relieve human suffering and restore health. At the same time, it is imperative that, in this biotechnologic age, we expand our moral imaginations to account for and be accountable to marginalized persons and concern ourselves with the shaping of a just future. Power is to be exercised on behalf of the least of us today and for the seven generations to come. If we wisely engage in shaping the future, we will create a world few of us ever imagined.

## Notes

1. The term "reprogenetics," coined by Lee Silver (1997, 8), underscores the increasing convergence of reproductive and genetic technologies. This convergence is particularly evident with regard to stem cells as evidenced by the recent announcement by Celera Genomics and Geron Corporation of a "... collaboration for human pluripotent stem cell genomics" (PE Biosystems 2000). Coupling genetic, nuclear transfer and stem cell technologies will potentially provide powerful tools for preimplantation genetic profiling, hES cell alteration, and germ line therapy.

2. The phrase "biotech century" is taken from the title of a book by Jeremy Rifkin (1998).

3. "Social lottery" refers to the manner in which one's social starting place affects opportunities. (Buchanan 1995). Here the term is used particularly with respect to class and socioeconomic standing.

4. Whereas embryonic stem cells are the current focal point, there is evidence that the more differentiated stem cells (AS cells) in the adult may be able to "switch fates." Bjornson and colleagues (1999) reported that mouse neural stem cells that give rise to three types of brain cells can also develop into blood cells when transplanted into mice whose bone marrow has been destroyed. Human mesenchymal stem cells were isolated from adult skeletal muscle and were capable of differentiating into multiple mesodermal phenotypes including skeletal myotubes, bone, and cartilage (Williams et al. 1999). In addition, stem cells from adult mouse skeletal muscle have a "remarkable capacity" to differentiate into blood cells including T and B cells (Jackson et al. 1999). If human stem cells derived from adult donors are consistently able to be channeled into particular cell and tissue types, they may be a viable therapeutic alternative to hES cells.

5. A subsidiary concern beyond the scope of this discussion is the potential for an increase in risky behavior by those with access to "replacement parts."

6. The Geron Ethics Advisory Board (1999), to their credit, set forth the principle that all stem cell research "must be done in a context of concern for global justice." Whereas the board's broaching of the justice question is crucial and courageous, it is difficult to fathom that Geron's primary obligation to its stockholders would be trumped by concern for marginalized stakeholders. In addition, meeting the demands of global justice is a complex matter that demands redistribution of resources well beyond the scope of a single company or a single medical advancement. For example, an argument from global justice could claim redirection of biotechnologic resources to the provision of clean water and sufficient food. For a further discussion, see Cahill (1999).

7. This responsibility for seven generations was expressed by Canadian aboriginals in their testimony to the Canadian Royal Commission on New Reproductive Technologies in 2000. In my usage, seven is to be seen for its symbolic meaning of completeness and totality.

## References

Bjornson, C. R. R., Reynolds, B. A., Magli, M. C., and Vescovi, A. L. 1999. Turning brain into blood: A hematopoietic fate adopted by adult neural stem cells in vivo. *Science* 283: 534–537.

Buchanan, A. 1995. Equal opportunity and genetic intervention. *Social Philosophy and Policy* 12: 105–135.

Cahill, L. S. 1999. The new biotech world order. *Hastings Center Report* 29: 45–48.

Cahill, L. S. 2000. Social ethics of embryo and stem cell research. *Women's Issues in Health* 10: 131–135.

Cho, M. K., Magnus, D., Caplan, A. L., McGee, D., and the Ethics of Genomics Group. 1999. Ethical considerations in synthesizing a minimal genome. *Science* 286: 2087.

Clarke, D. L., Johansson, C. B., Wilbertz, J., Veress, B., Nilsson, E., Karlström, H., Lendahl, U., and Frisén, J. 2000. Generalized potential of adult neural stem cells. *Science* 288: 1660–1663.

Geron Ethics Advisory Board. 1999. Research with human embryonic stem cells: Ethical considerations. *Hastings Center Report* 29: 31–36.

Jackson, K. A., Mi, T., and Goodell, M. A. 1999. Hematopoietic potential of stem cells isolated from murine skeletal muscle. *Proceedings of the National Academy of Science of the USA* 96: 14482–14486.

National Bioethics Advisory Commission. 1999. *Ethical Issues in Human Stem Cell Research*. Rockville, MD: National Bioethics Advisory Commission.

Parens, E. 2000. Embryonic stem cells and the bigger reprogenetic picture. *Women's Issues in Health* 10: 116–119.

PE Biosystems. 2000. Press release. Celera Genomics and Geron Corporation announce collaboration for human pluripotent stem cell genomics. www. pecorporation.com/.June 12.

Rifkin, J. 1998. *The Biotech Century*. New York: Jeremy P. Tarcher/Putnam.

Shapiro, H. T. 1999. Reflections on the interface of bioethics, public policy, and science. *Kennedy Institute of Ethics Journal* 9: 210.

Silver, L. M. 1997. *Remaking Eden*. New York: Avon Books.

Theise, N. D., Nimmakayalu, M., Gardner, R., lllei, P. B., Morgan, G., Teperman, L., Henegarie, O., and Krause, D. S. 2000. Liver from bone marrow in humans. *Hepatology* 32: 11–16.

Trauer, C. A. 1999. Private ethics boards and public debate. *Hastings Center Report* 29: 43–45.

U.S. Department of Health and Human Services. 1999. *Draft National Institutes of Health Guidelines for Research Involving Human Pluripotent Stem Cells*. Rockville, MD: National Institutes of Health.

White, C. B. 1999. Foresight, insight, oversight. *Hastings Center Report* 29: 41–42.

Williams, J. T., Southerland, S. S., Souza, J., Calcutt, A. F., and Cartledge, R. G. 1999. Cells isolated from adult human skeletal muscle capable of differentiating into multiple mesodermal phenotypes. *American Surgeon* 65: 22–26.

# 19

## Leaps and Boundaries: Expanding Oversight of Human Stem Cell Research

Cynthia B. Cohen

The demands of science and those of ethics with regard to oversight of stem cell research seem to be on a collision course. Groups within the scientific community, citing the promise of this research, argue that stem cell investigators should be allowed to pursue it unfettered by new forms of oversight. Additional review of this research, they believe, is unnecessary, would slow its progress, and would impinge on the value of scientific freedom (American Association for the Advancement of Science [AAAS] 1999). In contrast, others in the fields of ethics and public policy believe that if the research proceeds without special review, important ethical and social values will be overlooked in the race to attain its fruits. Indeed, some among them would halt this research altogether if it involves the use of human embryos (Doerflinger 1999). Others would require it to be reviewed within the current system of local institutional review boards (IRBs), but according to distinctive guidelines for this research (Wadman 1999; Weiss 1999b). Still others would establish a national panel to oversee stem cell research before it is addressed on the local level (National Bioethics Advisory Commission [NBAC] 1999, 102–105).

Are ethical and public policy considerations raised by stem cell research and its potential applications so weighty that this research requires national review? And, if so, what form should this review take?

Stem cell research requires special oversight not only because it raises important ethical and social questions of public concern, but also because it will inevitably converge with several other publicly sensitive technologies related to human reproduction. These include in vitro fertilization (IVF), cloning in the form of somatic cell nuclear transfer

(SCNT), and germ line modifications. These technologies, in combination, have potential to affect not only human well-being but the very meaning of what it is to be human. We should not, as a society, simply accommodate ourselves to pressing ethical and public policy issues created by stem cell research and related technologies as they appear willy-nilly. Instead, we should engage in anticipatory discussion of these issues and provide open national review of proposals (Cohen 1997).

These goals can best be achieved by establishing a multidisciplinary national panel whose purview extends beyond stem cell research to IVF, SCNT, and germ line modifications insofar as these converge with stem cell research. Moreover, this panel should have the authority to review research proposals in both public and private sectors and to recommend public policy for this innovative area of scientific endeavor.

### The Promise and Peril of Stem Cell Investigations

Research into stem cells offers major therapeutic benefits to humankind, many scientists maintain. They believe that these cells could be used to repair or replace cells damaged by disease, thereby alleviating the impact of such ailments as heart disease, diabetes, leukemia, Alzheimer's disease, and Parkinson's disease. Eventually, they predict, whole transplantable organs will be created from stem cells, reducing the risk of graft-versus-host disease. Scientists also envision using the cells in efforts to modify genes associated with certain genetic disorders. Research will also open new areas of basic scientific understanding, increasing our comprehension of human development and of factors that contribute to birth defects and infertility. Although scientists' enthusiasm about these potentials was dampened recently by hurdles encountered in the initial stages of the research, many believe that investigations will eventually prove successful (Weiss 1999c; Regalado 1998).

Whereas stem cell research offers tremendous promise, it also raises serious ethical and social questions. The issue that has received the most publicity is whether respect for the beginning of human life allows the use of embryos or fetuses. This is an extremely important and difficult matter that should not be dismissed lightly (Cohen 2000b). Yet it has overshadowed several other significant ethical and public policy issues that must be addressed wherever stem cell research proceeds—in the public or in the private sector.

### Issues of Safety and Efficacy

The possible use of stem cells derived from excess embryos remaining after IVF and from fetuses remaining after abortion raises questions related to risk and effectiveness.

No clear-cut criteria have been developed by which to measure whether these embryos and fetuses are free from disease and abnormalities. Consequently, concern is raised as to whether they can be adequately screened before they are used in humans. Furthermore, record keeping at some infertility clinics is not systematic and would not allow potential recipients of stem cells from surplus embryos to be assured that these embryos are within the range of normal. At least one investigator decided against using available excess embryos because he judged that they were of poor quality (Regalado 1998). Similar concerns arise about the condition of fetuses derived from elective abortions.

In addition, some scientists contend that the procedures required for implanting stem cells into the human body put certain patients at considerable risk (Smaglik 1999). Questions also were raised about whether these cells will grow normally once inserted into the human body. For example, stem cells derived from human embryos might create either benign or cancerous tumors in human recipients, as occurred in mice (Solter and Gearhart 1999). Research shows that when mouse germ cells are implanted into early mouse embryos, tissues containing these cells develop abnormally (Steghaus-Kovac 1999). There is no reason to think that human germ cells will not suffer the same difficulty. Scientists also conjecture that stem cell transplants from surplus embryos remaining after IVF would be rejected by the immune systems of recipients (Weiss 1999a). These possible negative effects of stem cells raise serious concerns about the risks of their application to human beings.

If federal funding is approved for stem cell research, this research should not proceed unless strong evidence shows that these resources can be used safely in human beings. Yet many local IRBs will not understand the scientific aspects of this complex cutting-edge work and may not have the time and support necessary to develop such expertise.

Furthermore, individual IRBs offer highly variable interpretations of federal guidelines; what is accepted by one institutional IRB may be rejected by another. Some local IRBs were cited by the Office for Protection from Research Risks for failure to conduct appropriate and timely reviews of research (Foubister 1999), indicating serious flaws in the way that some of these bodies function. Furthermore, IRBs have limited means of ensuring that researchers follow their recommendations after their protocols are approved. These local bodies, moreover, are specifically prohibited from considering long-term effects of biomedical research on ethical and social values. Consequently, they could play little role in assuring the public that these values were taken into consideration during their reviews. Finally, IRBs are subject to conflicts of interest if their decisions about potentially profitable research would have a negative impact on the financial health of their institution. This can make it difficult for them to retain their impartiality and objectivity.

The alternative of moving to the federal level for oversight would not remedy these problems under the current system of review. The Food and Drug Administration (FDA) is authorized only to address issues of safety and efficacy when human beings are direct subjects of experimentation. Thus, it is not empowered to oversee stem cell research. Some other group is needed to provide national review of the risks and effects of using stem cells in the early stages of research if the public is to be assured of its safety.

### Issues of Consent, Pressure, and Coercion and Possible Overproduction of Embryos

Women who undergo abortion and those who are infertile find themselves in circumstances whose personal, medical, and social dynamics raise questions about whether they may be subject to undue pressure or coercion (National Institutes of Health [NIH] 1994). Federal regulations governing the donation of fetal tissue derived from induced abortions (42 U.S.C. Sec. 289g–l) address issues of free and informed consent to a certain degree as these relate to donors of fetal tissue. The situation of women who might donate excess embryos for stem cell research, however, is even more complex and raises the possibility that they will suffer undue stress and even coercion in the process.

The meaning of infertility is shaped by the ideology and social structure of a society, by its beliefs about the importance of blood relationships, its role expectations of men and women, the social value it ascribes to children, and its medical ethos and technologic capabilities (Greil 1991). Infertile women, Greil finds, are viewed as outcasts in our society. Therefore, in the United States, infertile women undergoing IVF are advised to produce and implant as many embryos as possible within what is known about the limits of safety so that they can have an improved chance of creating children and thereby making a place within our society. Many infertile women agree to undergo procedures that are onerous and risky because they are convinced that this is necessary if they are to conceive (Bartholet 1993). Infertility specialists are also under pressure to produce as many embryos as they can for women or couples while avoiding hyperstimulation that could lead to patient injury. The more embryos they create and implant, up to a point, the better the chances of success. It is extremely important for those in reproductive medicine to produce high rates of success if they are to do well in the competition for patients and maintain a reputation for excellence. Thus, potential embryo donors going through IVF are thrust into an intensely pressured atmosphere that borders on the coercive, even though no one has directly confronted them in a coercive way.

Because of this pressure, special attention is required not just to research using excess embryos, but also to the practice of IVF itself. Guidelines and standards are called for to ensure that donors of embryos are not coerced or pressured into creating extra embryos for stem cell research at the time of treatment, are not coerced or pressured into donating extra embryos at the end of treatment, and have the opportunity to give adequately informed consent both to the creation and donation of extra embryos. Draft guidelines governing the donation of excess embryos for stem cell research were developed by the special advisory panel convened by the director of the NIH (64 Federal Register 67576). They are very narrow in scope and do not address the practice of IVF or the range of ethical and social issues raised by the use of fetuses and embryos in stem cell research or IVF (Wadman 1999; Weiss 1999b).

Although IVF has necessarily been conducted in the private sector due to a ban on the use of federal funds for embryo research (P.L. 105–277, section 511, 112 Stat. 2681–386), it has not had the sort of scientific review and verification afforded other forms of human experimentation that have received federal funding. No IRBs regularly review IVF research protocols; the FDA does not regulate research in this area; and there is a dearth of careful clinical trials of IVF and other reproductive technologies. The Clinical Laboratories Improvement Act (42 U.S.C. Sec. 263a) requires that laboratories performing tests related to assisted reproduction must be certified for quality assurance. However, it does not cover such techniques as handling gametes or embryos or the IVF process itself. Thus, IVF gradually moved from the experimental stages into practice without consistent scientific and ethical review accessible to the public.

Although the American Society for Reproductive Medicine (ASRM) tried to step into this gap, its guidelines for research and practice are not binding and there is no way to ascertain whether practitioners are following them. These guidelines offer important considerations that should be taken into account in gaining informed consent from donors for the use of embryos for research (ASRM 1997), but they also create certain ethical concerns. For instance, they allow infertility specialists to request embryos that might well benefit their own research, thereby creating a serious conflict of interest. One guideline of the NIH advisory panel stipulates that the decision to create embryos for infertility treatment must be made separately from the decision to donate embryos for stem cell research. It does not, however, avert the heightened pressure to produce multiple embryos that patients experience during IVF treatment (Cohen 2000a).

These considerations underscore the need for some form of review of the creation of embryos during IVF and the use of excess embryos remaining for stem cell research. A standing national panel is necessary to pay continuing attention to ethical concerns about consent, pressure, coercion, and overproduction of embryos in the context of IVF. If the use of excess embryos is ultimately not approved for federally funded stem cell research projects, it will still be necessary to oversee their use in private sector, as serious ethical and public policy questions are raised by embryo production whether it is conducted in the private or public sector.

**Ethical Issues Related to SCNT and Germ Line Interventions**

As stem cell research proceeds toward human trials, it will necessarily include the use of embryos created by means of SCNT. Inevitably SCNT will have to be applied to overcome the difficulty that stem cells derived from excess IVF embryos and cadaveric fetuses would likely be rejected by the immune systems of recipients. The procedure would allow autologous cell replacement with stem cell lines derived from the patient's own cells.

Private industry, which currently controls the direction of stem cell research, plans to introduce SCNT into its investigations. Roslin Bio-Med, a Scottish corporation created by the scientists who created the first cloned sheep, recently entered into an agreement with Geron in the United States, which funded early stem cell research. By matching its stem cell technology with Roslin's cloning technology, Geron plans to produce cells, tissues, and eventually whole organs for individual patients.

The use of SCNT raises significant ethical and public policy concerns because it is a form of cloning. It is possible to implant embryos produced by this method into the human uterus to produce new human beings. In testimony before NBAC in January 1999, Gearhart and Solter maintained that SCNT can be confined to the embryonic level. Others, however, believe that it would not end with the destruction of cloned embryos, but would pave the way for the first births of cloned babies (Weiss 1999a, A1).

Another controversial procedure that will converge with stem cell research is an intervention made into the human germ line that changes the genetic makeup of a line of future individuals. Also testifying before NBAC, Austin Smith suggested that stem cells may prove useful in human gene transfer experiments aimed at treating genetic disorders. Because such cells have the capacity to proliferate over a period of time (Thomson et al. 1998), it is easier to make precise gene insertions into them than it is into other kinds of cells. This means that it would be easier to create inheritable genetic modifications in stem cells than in differentiated cells (Parens 1999). Another way to try to overcome rejection of implanted stem cell lines would be to create approximately twenty immunologically different lines of embryonic stem cells, which would allow immune compatibility with most of the population (Wadman 1999). One way of avoiding the need to create these cell lines would be to alter the genes that control the major histocompatilibility complex (Fletcher 1999). This, however, would reopen the door to controversial germ line interventions.

The possibility that SCNT and germ line interventions could be used as stem cell research develops raises weighty questions. Until some form of public review and oversight of this research is established, safeguards are required against transferring an embryo created by SCNT into a human uterus. The NIH advisory panel recommended that regulations be written prohibiting implantation of cloned embryos into the uterus in federally funded research projects. However, no federal agency is currently charged with continued review of stem cell research after it has received IRB approval to ensure that such a prohibition is enforced. Moreover, such regulations would not bar private investigators from engaging in human cloning.

Public discussion of the prospect of intervening into the germ line is necessary. The FDA, with the advice of the Recombinant DNA Advisory Committee (RAC), has oversight and review authority over any proposed therapeutic modifications to recombinant DNA in gametes or embryos. However, the FDA does not have standing to engage in open, interdisciplinary discussion of the ethical and social implications of new treatments, nor does it have a mandate to encourage open national debate on controversial issues raised by significant scientific studies or to seek national consensus about them (Cook-Deegan 1999). Quite the reverse: the FDA usually meets behind closed doors, leaving the public with no direct means of verifying that it has carried out its reviews appropriately and has carefully monitored research involving human subjects.

The RAC, in its earlier years, reviewed not only federally funded research protocols involving recombinant DNA, but also private sector protocols sent to it voluntarily. In effect, it operated as a national IRB for research into recombinant DNA. It also functioned as a national ethics advisory board for ethical and policy issues related to research into recombinant DNA, asking for public deliberation and input from many constituencies. The role of the RAC as a national review board ended in

1996 (Marshall 1996). It currently reviews "novel" protocols related to recombinant DNA research. Unfortunately, it has only symbolic authority to deal with such protocols and cannot prevent those it considers questionable from proceeding. The RAC could be called upon, if special authorization were passed, to examine any stem cell research proposals that incorporate interventions into the human germ line. Such authorization has not been passed. Because its purview is restricted to research involving recombinant DNA, the RAC has no standing to review protocols that incorporate IVF or cloning. Thus, no forum currently exists in which the ethical and social questions raised by stem cell research and the technologies with which it is intimately interconnected can be addressed in a public and coordinated way today.

## Forms of Oversight of Stem Cell Research Recommended to Date

What body should establish guidelines and standards for stem cell research and ensure that safeguards are applied consistently by scientific investigators? The report of the AAAS (1999) maintains that the ethical and policy concerns raised by investigations in this area are "not unique to stem cell research" and that no special review of this form of research is necessary. Yet serious inadequacies of current methods of oversight in this sensitive area have been noted above. Reports that scientists and drug companies failed to notify the NIH of deaths that occurred during gene therapy experiments, contrary to federal reporting requirements (NIH not told 1999), strongly indicate that research in publicly sensitive areas has to be subject to stringent oversight at the federal level in a publicly available manner.

Three other bodies have maintained that some form of review is necessary. The NIH considered it essential to set out new rules governing stem cell research, but left review in the hands of local IRBs (Brainard 1999). Depending solely on local IRBs to review these investigations, however, raises difficulties noted above. Moreover, the NIH disbanded the ad hoc panel that it convened to develop criteria for reviewing stem cell protocols. This leaves no mechanism for further development and modification of guidelines in light of investigators' experiences. A standing body is needed to address new issues as they arise.

Geron Corporation established a standing ethics advisory board to review its research, maintaining that IRB review alone was insufficient (GEAB 1999). Other private biotechnology groups, unlike Geron, established no ethics review panels. They work in greater secrecy, engaging in little open discussion of ethical questions surrounding the work (Regalado 1998). No national body is available to which stem cell investigators in the private sector can turn for review in the way that many private corporations voluntarily turned to the RAC.

The NBAC (1999, 102–105) urged that a standing federal oversight panel be established for stem cell research carried out with federal funds. The need to assure the public that the research is being undertaken safely and according to ethical guidelines was foremost among its considerations. The NBAC recommendations for a national stem cell oversight and review panel follow in certain significant respects the model created by the RAC. The panel would review protocols and monitor research as it is carried out across the country, providing some assurance that investigators were adhering to guidelines and standards. The panel would also keep track of the history and use of stem cells and produce periodic reports on the current status of research. Furthermore, it would serve as a resource for guidance regarding ethical and social issues. In the NBAC model, private groups would be encouraged, but not required, to present their protocols to this national panel for review.

## The Need for an Oversight Body with a Broader Purview

As valuable as NBAC's proposal is, it does not go far enough. The commission took an incremental approach to the approval of stem cell research, allowing resources and methods to be used that would be required in its initial stages (Fletcher 1999). It did not go further to consider ethical concerns that would arise when research reaches the point where it must employ SCNC and germ line interventions to avoid rejection of stem cell transplants and to modify genes in stem cells associated with genetic disorders. Evidently, NBAC felt that it did not have to resolve the larger questions if it could address those immediately at hand satisfactorily.

The NBAC's recommendation would ultimately lead to creation of a string of distinct ethics review boards for research involving stem cells,

IVF, SCNT, and germ line interventions. Maintaining a collection of such panels is likely to result in slow reviews and incompatible recommendations. Moreover, the purview of the panel would not go beyond research efforts to include aspects of IVF that are closely associated with the use of embryos in stem cell research. Public review of emerging applications of these transformative biomedical technologies and of reproductive practices with which they are closely associated should be carried out in a more fully coordinated way.

A National Stem Cell and Associated Technologies Advisory Board (NSCATAB) should be established in the United States under the aegis of the U.S. Department of Health and Human Services. Its purview would extend not only to stem cell research, but also to the ethically sensitive technologies with which it is necessarily associated, including IVF, SCNT, and germ line interventions. Its goal would be to ensure that the research and related technologies are carried out safely and ethically and that public policies are well informed. Such a board would provide a flexible and long-term oversight mechanism to address expected and unexpected developments and would also offer a forum for discussing pressing ethical and public policy questions. The board would have the authority not only to establish consistent guidelines for stem cell research and closely connected reproductive practices, but also to modify these as circumstances change. The history of the RAC suggests that national oversight is effective and that it not only provides thorough scientific and ethical review, but encourages national discussion and debate.

The NBAC called on privately funded investigators voluntarily to submit their research protocols to the oversight panel that it recommended. However, those carrying out research in the private sector inevitably face a conflict of interest: despite the example set by Geron, the drive to remain competitive pushes them into secrecy. Those in private industry will require strong motivation to ask voluntarily to have their protocols reviewed by the NSCATAB. Some will do so to advertise that they have received the imprimatur of the national board. Others, however, will not.

Scientific freedom and corporate productivity are important goods, but they do not outweigh the public's need to be assured that experimentation involving humans is safe and that it honors ethical and social values that are significant to our society. To provide such public reassurance, it is important to unify oversight of publicly and privately funded stem cell research in a national body. Under such a system, private corporations would be able to keep proprietary information hidden, as they did when the RAC reviewed recombinant DNA proposals from the private sector, yet still submit their research protocols for review by NSCATAB. Even if Congress prohibits the use of fetuses and embryos in the pursuit of stem cell research, it is imperative that the private sector undergo some sort of review of its work.

It will take time to develop a standing national board to assess innovative protocols related to this research and the technologies with which it is associated. Legal requirements for rule making would have to be met before a new body could be created. This would delay the start of innovative research for a significant period of time. Therefore, until a national review board is established outside the provenance of the NIH, the current NIH stem cell advisory panel should be reassembled and the diversity of its membership expanded. Moreover, its charge should be revised to enable it to review all stem cell research protocols.

In summary, a public oversight body is required that will monitor this work as it is carried out across the country. This body would also prepare for the prospect that significant issues of public concern related to the use of cloning and germ line interventions will have to be addressed.

Stem cell technologies bear the potential for remaking the very beings that create them. Thus, this emerging area of research raises deeply important ethical and public policy issues that must be addressed in a coherent, principled, and open fashion. A method of oversight can be adopted that offers considerable freedom to publicly and privately funded investigators while at the same time protecting human subjects against harm and society against diminution of our basic social values.

## References

American Association for the Advancement of Science and Institute for Civil Society. 1999. *Stem Cell Research and Applications Monitoring the Frontiers of Biomedical Research*. Washington, D.C.

American Society for Reproductive Medicine. 1997. Informed consent and the use of gametes and embryos for research. *Fertility and Sterility* 68: 780–781.

Bartholet, E. 1993. *Family Bonds: Adoption, Infertility, and the New World of Child Production*. Boston: Beacon Press.

Brainard, J. 1999. U.S. bioethics panel urges Congress to lift ban on stem-cell research. *Chronicle of Higher Education* September 24, A46.

Cohen, C. B. 1997. Unmanaged care: The need to regulate new reproductive technologies in the United States. *Bioethics* 11(3/4): 348–365.

Cohen, C. B. 2000a. The use of "excess" human embryos for stem cell research: Protecting women's rights and health. *Women's Health Issues,* vol. 10.

Cohen, C. B. 2000b. Open possibilities, close concerns: The import of religious views on the future of stem cell research. *Park Ridge Center Bulletin* 13: 11–12.

Cook-Deegan, R. 1999. Address to the American Association for the Advancement of Science public meeting on stem cell research, August 25.

Doerflinger, R. M. 1999. The ethics of funding embryonic stem cell research: A Catholic viewpoint. *Kennedy Institute of Ethics Journal* 9(2): 137–150.

Fletcher, J. 1999. Deliberating incrementally on human pluipotential stem cell research. National Bioethics Advisory Commission, background papers on embryonic stem cell research.

Foubister, V. 1999. More centers cited for ethics lapses in research. *American Medical News* 42(41): 8, 10.

Geron Ethics Advisory Board. 1999. Research with human embryonic stem cells: Ethical considerations. *Hastings Center Report* 29(2): 36–38.

Greil, A. L. 1991. *Not Yet Pregnant: Infertile Couples in Contemporary America.* New Brunswick, NJ: Rutgers University Press.

Marshall, E. 1996. Varmus proposes to scrap the RAC. *Science* 272: 945

National Bioethics Advisory Commission. 1999. *Ethical Issues in Human Stem Cell Research.* Vol. I. Rockville, MD: National Bioethics Advisory Commission.

National Institutes of Health, Ad Hoc Group of Consultants to the Advisory Committee to the Director. 1994. *Report of the Human Embryo Research Panel.* Vol. I. Bethesda, MD: National Institutes of Health.

NIH not told of deaths in gene studies. 1999. *Washington Post* November 3, A1, A6–7.

Parens, E. 1999. What has the President asked of NBAC? On the Ethics and Politics of Embryonic Stem Cell Research, NBAC background papers on embryonic stem cell research.

Regalado, A. 1998. The troubled hunt for the ultimate cell. *Technology Review* (July/August) www.techreview.com.

Smaglik, P. 1999. Promise and problems loom for stem cell gene therapy. *Scientist* 13: 14–15.

Solter, D. and Gearhart, J. 1999. Putting stem cells to work. *Science* 282: 1468.

Steghaus-Kovac, S. 1999. Ethical loophole closing up for stem cell researchers. *Science* 286: 31.

Thomson, J. A., Itskovitz-Eldor, J., Shapiro, S., Waknitz, M., Swiergiel, J., Marshall, V., and Jones, J. 1998. Embryo stem cell lines derived from human blastocysts. *Science* 282: 1145–1147.

Wadman, M. 1999. NIH stem-cell guidelines face stormy ride. *Nature* 551: 185–186.

Weiss, R. 1999a. Embryo work raises specter of human harvesting, medical research teams draw closer to cloning. *Washington Post*, June 14, A1, A4.

Weiss, R. 1999b. Panel drafts ethics plan for embryo cell studies; rules would guide federally funded research. *Washington Post* April 9, A2.

Weiss, R. 1999c. Stem cell discovery grows into a debate. *Washington Post* October 9, A1, A8–9.

# Jordan's Banks: A View From the First Years of Human Embryonic Stem Cell Research

Laurie Zoloth

When the first serious work in genetics became possible, the initial public reaction to the exploration and manipulation of the genetically coded structure of the human being and to the research on that code was a curious mixture of fascination and fear. Fascination drove an intense public interest in each new genetic advance, project, or claim, and support for research projects that appear to have the potential for therapeutic use. Fear prompted both initial caution and legislation to limit that research. Limits on the clinical use of genetic interventions and on research and testing were created for four salient reasons.

First, technical barriers made successful manipulation of genetic material for reliable medical use highly risky. Hence, regulators first focused on issues of safety and avoidance of the clearly foreseeable chances for harm, understanding that even with the best of intentions, unforeseen error and unintended consequence were unavoidable.

Second, the mere activity of intervention into human DNA was viewed with alarm, and scientific use of human embryonic tissue as the venue to explore and manipulate human DNA was seen as a violation of essential moral limits and of the nature of the meaning of the self. Hence, political and legal "bright lines" were erected to prohibit certain experimental trajectories, and political and religious pressure was exhorted to prohibit intellectual discourse in various aspects of the field that so dramatically raised the question of human mortal limits and fears about playing God.

Third, the problems of informed consent and refusal that are raised by all sorts of medical research seemed especially daunting in this research, surrounded by both great hopes and great uncertainties. All eggs and sperm and embryos used in this research have to come from consenting donors, usually from participants in vitro fertilization (IVF) or abortion clinics, who are often in the midst of a protracted infertility or fertility crisis themselves.

Finally, the specific problem of destruction of embryos required for much of this research created a discursive arena directly proximate to the deepest moral divide in American political life, the abortion debate. This debate has served as the great simile in several crucial medical ethics controversies, a litmus test for our collective understanding of women, sex, faith, and death. Nowhere is the impact more clearly felt than in questions surrounding genetic research. Given this, public funding sources available to researchers were limited, curtailing active searches for new techniques in this area. In particular, research on human embryos is limited by federal bans on funding, and research on the manipulation and alteration of germ line DNA for therapeutic medical purposes is constrained by federal law.

But the enormous potential of such intervention is a powerful incentive, and as the practical technical skills of genetic scientists improve, ethical issues at the margins of the research are again raised for reconsideration. Private companies quietly continued to fund university researchers, and work on the human genome and embryonic cellular manipulation continued. In fact, research in human embryonic stem cells (called "The search for the Tabula Rasa of genetic research," or "the Ultimate Cell"; (Regalado 1998) proceeded swiftly, and breakthroughs in this technology again raised questions about the ethical implication for such interventions in the clinical context.

Details of genetic science necessary to understand these cells, and the potent possible applications of human embryonic stem (hES) and human embryonic germ (hEG) cells are the subjects of other chapters in this volume. This chapter addresses ethical issues that emerged in the first considerations of this technology, what many of us in the field of bioethics were deliberating as we learned of the new science and confronted the ethics issues it raised. Colleagues were asked to reflect on early stages of the research who were members of institutional review boards (IRBs), and the Geron Ethics Advisory Board (GEAB), and the National Bioethics Advisory Committee (NBAC) as the field debated the issues of consent, moral status, use of animal tissues, abortion, use of fetal tissue, and the nature and goals of entrepreneurial research. In this new capacity, ethicists weighed the problem of privacy, the role of justice considerations, and issues of the marketplace in science. At the onset, I want to be clear that far more issues remain unresolved than are settled, that the territory ahead is largely unexplored, and that the single most important task that faces us as a field is a steady call for continuing conversation and public debate.

## The Journey so Far: Reading the Maps to the West

Ethics begins in the narrative and the case. Early in 1998 several of us who had published in the field of genetics and ethics were approached by the leadership of Geron to participate in reflection on their work. Aware of deep political and theological controversies in the field, and after having watched the media frenzy that followed the announcement of mammalian cloning, Geron executives wanted to understand and to consider emerging thinking on ethical issues of the new genetics. The scientist from Geron explained the concept of stem cells to this philosopher by drawing on a napkin the picture of what they were trying to accomplish, and as the implications for the work began to be explained, the medical uses of such a technology were immediately apparent and extraordinarily exciting. These cells, he told me, were "magic." This interaction set the stage for the direction and structure of much of our subsequent work, which began in earnest six months later.

The Geron board was introduced formally to the notion of the hunt for hES cells with a careful scientific explanation of the mechanisms of the research and the motivation for the science itself. In this, we wanted to understand the exact physical potential, capabilities, and biology of the early embryo. We decided that that we would have to focus on three issues: (1) *telos*, including both practical medical ends and speculative but foreseeable correlative ends as they affected our ontologic notions of person, aging, and death; (2) *process*, and the question of origin, derivation, power, special concerns for women, and consent (beloved of ethicists); and (3) *context*, including justice, commodification, and implications of the work.

But like all philosophic discourse in the first few months of the debate, we had to learn some intricate science and we had to do it very rapidly. We learned of the success of the procedure we had just heard about and suddenly were in the midst of crafting our responses to it. We were not unusual in this: IRBs that were consulted at universities in which the work was housed and later, NBAC, National Institue of Health (NIH), and American Association for the Advancement of Science (AAAS) committees followed the same path. We all learned the technical terms for stages of embryonic development, the status of embryos in IVF clinics from which they were obtained, and that many were discarded and often donated or stored. This link between derivation, reproduction, and hence consent occupied the field from the onset. Since the IVF clinic is such a vexed locale for bioethicists, we were drawn into debates over that setting. Hence, many of our first discussions focused on the informed consent form, the issue of what constituted a parent in this case, and the emotive quality of the experience, wondering if real informed consent was even possible in that setting.

A second focus of concern that arose early was the use of embryos. Our understanding was that embryos were graded (1–4) with grades 1 and 2 considered useable for implantation, and grades 3 and 4 considered too physically imperfect to be used. It was explained in the first discussions that no use could be made of stem cells in the reproductive process, that the cells could never implant in a uterus and develop into a fetus in any case, and that this technology was neither a germ line intervention nor did it rely on controversial techniques such as cloning for human reproduction. We understood at that time that no embryos would be created for the purposes of research, that embryos used would be donated under the most stringent system of informed consent, and that these embryos would have been discarded in any case.

Armed with this information, the GEAB turned toward precedents and began an intricate search through source faith traditions, normative philosophic texts, and prolonged debates on issues of a just contract in research and informed consent (to address consent), and on issues of treatment of the person who is doomed to death (to address embryo destruction). We put aside the question of moral status since we understood ourselves to be reflecting on the use of embryonic tissues with no possible future; poorly graded embryos, because they were physically imperfect, would never be used to create children. For this, there is a rich literature of casuistry in many religious traditions. For many faiths, when one is inevitably dying, our obligation to save life, especially relative to other lives, is altered. One could argue that perhaps our civil obligation toward these doomed embryos could be understood in this manner. The idea that the embryos were not viable, and that their use allowed their genetic material to be carried on in some way, made logical analytic sense. The notion that use could be made of microscopic tissues that would otherwise not even be frozen for future use seemed on all accounts to be a beneficent one. It is an argument that still can be found in many articles on stem cells.[1]

## Beyond Known Borders

In the next few months the science rapidly moved forward,[2] and hence our elaborate rationale, although certainly interesting, became archaic. Since the norm for implantation was now two instead of five, many more embryos were available but would not be used by a particular couple, giving rise to a new ethical problem of "spare" embryos and their use and storage. This raised a host of new concerns: could embryos be used by other couples, and if so, to whom did they belong?[3]

Hence, after this understanding, it became apparent that a critical issue was moral status of these embryos, and that moreover, we had to pay attention to the use of animal embryos in growth medium that supported the tissues. Here, too, we did more textual research and found that in several faiths the unimplanted, noncorporeal embryo was not regarded as morally equivalent to a human life. In fact such research might well be mandated to save life in Jewish law.[4] At NBAC hearings held around this time, one could hear diametrically opposing arguments about what each religion mandated by its legal codes.

As we openly reflected on this issue in public forums, we heard of other ethical and scientific considerations. For example, colleagues and other researchers drew our attention to the work of Nagy et al. (1993)

in which living mice were successfully gestated from a cluster of stem cells placed in a trophocytic matrix (to simulate placenta-forming cells), aggregated with tetraploid embryos, and placed into a mouse uterus. Thus in theory, if given the correct matrix, stem cells can make all parts of a living organism, at least in mice. If all stem cells could potentially become embryos given the right sort of cellular environment, what exactly does one have when one has a stem cell cluster? McGee and Caplan (1999) were led to ask: "what is in the dish?" Is it a canonical cell line? Or is it actually also a potential human fetus awaiting the right technology to transform it into being?

Later advances drove our response. As researchers became interested in exactly how cytoplasm and the nucleus worked their magic, how the reprogrammed cell was stimulated to differentiate, attention turned toward how this happened in animals. As Geron acquired an alliance with the Roslin Institute and scientists involved in mammalian cloning, we began to rethink our stance on animal cloning for research purposes. If we were even to understand the basic science of early embryology and of cellular function, would we not have to observe this reprogramming over and over? Meanwhile, at AAAS meetings, scientists assured philosophers that use of adult stem cells for medicine was neither possible nor practicable. Unlike cells of the embryo, which were totipotent, adult stem cells were limited in their capacity. This scientific understanding led our thinking as we crafted our moral response.

But the real potential of stem cells lies in their potential as sources for tissue and even eventually organ transplantation. Presently, two factors limit organ transplantation: availability and histocompatibility. The latter is a deeply complex problem.[5] In the word of one transplant surgeon, "after 20 years at this, I have come to see transplantation as exchanging one terrible chronic disease, say, heart failure, with another terrible chronic disease, graft-versus-host."[6] Many uses of stem cells rest on solving this problem. If the DNA of the stem cell was matched to the tissue of the recipient, or if a universal stem cell could be engineered, rejection would no longer be a problem. As this issue of engineered histocompatibility was explored, we focused on what issues this technology, which rests on transferring a DNA-matched cell nucleus to a stem cell cluster, would engender, and we were not alone.

## From Good Science to Good Ends

Suddenly the literature was full of this speculation. If stem cell technology develops in tandem with nuclear transfer technology (the first step in cloning), cells not only could potentially be programmed to differentiate into specific tissue types, but could be personally genetically tailored for each transplant recipient. It is a stunningly important technology, potentially eliminating graft-versus-host disease and saving millions of lives. However, it raises the problem of reproductive uses, moral status, and instrumentality. And if such nuclear transfer could be done, might not the DNA of the transferred nucleus could be manipulated as well, raising the complex problem of inadvertent or deliberate germ line intervention, yet another ethical issue on which our society will have to reflect. In the first stages of careful discernment, we carefully looked at each separate issue. We were at first like a committee knowing parts of an elephant and never knowing the whole beast. But now we could worry about everything at once; now we were beginning to see the whole creature. As Eric Parens persistently warned, we were beginning to have all of the issues we were worried about in bioethics in the same room: consent, women as research subjects, sex, IVF clinics, animals, abortion, germ line intervention, and aging.

Let us be clear here too—the more we in the field know, the more we care about the stakes of the research. Commissions include parents of children with diabetes; we hear pleas from patients frankly desperate for research.[7] At Geron, we were given white coats and taken to the lab where we saw beautiful new beating cardiac cells, the exquisite intricacy of newly differentiated structures. All who looked at the problem worried: if bioethicists stop this research and defeat the scientific vision, we might be doing an unthinkable moral wrong. Hence, in report after report, we ethicists were strongly supportive of nearly anything that was suggested for further exploration, even while we struggled with the implications of medicine transformed by this science.

In part, difficulty understanding the moral meaning of these implications is a problem of language, a language of discourse about human reproduction that emerges from classic understanding of gametes, families, and sexual reproduction, and debates about abortion, birth control, privacy, rights, and sanctity. What made much of this so heavily freighted in religious texts was the inescapable link among the body of a woman, the passion of the erotic, and the fact of a child; that is, the question of female control. However, as we contemplate a world of cellular replication and reproductive potential without gametes, we will require new language to describe what we intend, its moral meaning, and what we find fitting. We will require ways to understand not only the power relationships, but the way that the frankly miraculous—the developing embryo—is explained.

But the narrative of our work when told in this mythic way, with ethicists dazzled by the scientists,[8] raises, I believe, a far deeper question indeed, that of the goal, meaning, and obligation of ethics consultation at the frontiers of new technology. In this retelling, I think we have as yet failed to realize the full potential of our vision in bioethics. And in this way, much of what we have done to this point has been partial, inadequate, and underdetermined. At stake is not only the rules of play, and not only the consequences of our action, and not even the problem of the status and meaning of embryos, but the question raised by James Keenen (1999): *who are we and what do we become when we do this thing?*

I want to call for the next bold step in this arena and invite colleagues into something more than the discourse that the GEAB called for in the *Hastings Center Report* (Geron 1999). For in here or on other regulatory bodies, we will be reacting to what is asked of us, rather than setting the questions ourselves, and asking ourselves what ought to be our goals, and what this work will make us become. Ethicists have to think in a way that I will name "Exodic,"[9] meaning we must ask what *ought* to be the case, what would be exemplary research, rather than struggling to figure out if the "is" presented to us is acceptable or not.

## The Logic of the Good Consequence

Thinking of our narrative in the way I described, following rather breathlessly just behind the science, has meant that we are led inexorably into a path of beneficence-based utilitarian analysis. Something like this inevitably follows: A. Stem cells have extraordinary medical potential to save lives. If A, then B: all basic research in stem cells is critically important to explore its clinical possibilities.[10] If B, then C: to be useful for transplantation to particular human bodies, stem cells must be histocompatible. This question aside, it is essential to understand early cellular development better if we are to attempt to program embryonic stem cell into specificity or deprogram adult stem cells into simplicity. If C, then D: if we want to understand how to control the stem cells[11] or why mutations that may lead to cancer occur, or to control genetic characteristic of stem cells, we will have to observe the first steps of early cellular development. If D, then E: we must do experiments on early embryos (some of which are already being done in fertility research), limited only by the widely understood ban on allowing human embryos to develop beyond a fourteen-day limit.

But thinking along such lines immediately rapelles us down the slippery slope. For if we allow researchers to create embryos and use some of them for nonreproductive experimentation, we are forced into a logical crevasse, arguing details after the initial point is conceded. Young (chapter 15) maintained that hundreds of excess embryos are created by fertilization and destroyed in the process of traditional IVF. Of course, what exactly we mean by excess is rather slippery as well. How different is this from the entire mountaintop of ways we could manipulate embryos?

Once the destruction of embryos to create stem cells is approved (as in reports of GEAB, HERP, NBAC, AAAS) we find ourselves drawn deeper and deeper into ways that we have to study them, manipulate them, alter them. In short, once what is at stake is that we have concluded that the ends of medicine mandate these means, we can abandon other appeals. The power of the narrative is thus the shaper of the telos. We see, for example, shifts in the words that are used to describe the debate, as in newspaper accounts referring to the hES tissue as "tiny blobs" of cells, the repeated description of how microscopic they are. That this is technically true misses the point, as in some way does all debate about moral status or moment of ensoulment. What is at the heart of the issue is to ask: are some things so important to human advancement that we have a positive responsibility to pursue them? Who are we if we turn away, and who are we if we proceed?[12]

## The Search for Alternative Cases

It is the social construction of the language of the narrative that is the most seductive.[13] Bioethicists can easily become, first, mesmerized by the drama of the science (or rather the promise of the science) and then fall back on our familiar firewall position of consent, and become mesmerized by the complexities of figuring out our consent forms, and never fully develop the larger contextual reflections on justice, never really figure out what that means to us.[14]

Before the politics, the lobbying effort, the siren call of science and the grant application, we must ask ourselves as ethicists in a discipline with certain attachments, if not agreements on principle and case narratives, what is the good act and what makes it so? If we do not do this we will be committed increasingly to an inevitable consequentialism, and we will not hear the appeals from virtue, from deontology, or from fidelity. But there are many appeals other than utilitarianism and consequence. Consequence, as Callahan (2000) reminded us, and the struggle to protect research subjects, is just not a sufficient justification for research at all costs. One idea is to reflect on the relationships of family[15] or of a community in which a life is sacrificed for an overwhelming need. One such example is the just war theory and it is to that that I believe we might turn, for example, rather than to reproductive policy theories, which are faltering in analogy and enmeshed on the old battlegrounds. In just war theory we have a discussion of many comparable issues: why life can be taken in selected instances and what to do if members of a religious faith oppose war based on conscience. This discussion allows for three things. It gives full consideration to the Roman Catholic position that a life is being lost, without minimizing the respect that one has for life in its many and varied expressions and development. It allows for the notion of moral objection, resistance, and witness, and thus conscientious objection in the face of a wider social decision, rather than inaction or an attenuated social policy. The discussion has the appropriate quality of serious engagement for medical theory and research, rather than assuming that short-term policy will inevitably lead to normative practices of casual use of human embryos for frivolous pursuits. This position is not perfect—nothing is. Several contend that it is like a war in which innocents are targeted, and that just war theory cannot be used as an analogy. Other disagree, noting that the loss innocent life was always a feature of war.

## The Exodic Moment before Settlement

What we need is a vision that looks even beyond the horizons that keep appearing before us as the science develops. We have to develop both the moral imagination and ethical casuistry that allows creation of norms and standards for our work in a new terrain. Here is what I mean by "exodic" thinking: can we halt for a moment, before we cross critical dividing lines, and ask ourselves and researchers: under what banner do we lead? For this moment two metaphors come to the mind of this Jewish ethicist. The first is a biblical one about the nature of sin and the nature of responsibility inherent in the taking of a new terrain. It is the Hebrew people, a generation after Exodus, after exile, first entering the land that promises so much and is after all, promised, but that can be destroyed if the community is not aware of the daily conundrum of the choice before it, and not aware that the problem faced is not geographic or exploratory, but an ethical choice:

See! I am setting before you today: blessing and curse. (Deut. 11:26–29)

I call heaven and earth to witness against you this day: I have put before you life and death, blessing and curse. Chose life—if you and your offspring would live—by loving the Lord your God, heeding the commandments, and holding close. (Deut. 30:19–20)

The Hebrew wanderers are assembled at the very gates to the long-promised land and Moses speaks to them, opening with a word: See! *Re'ah*! Behold![16] Moses is speaking to the people of the world that they will encounter a land unseen, in a place and time that are distant. The choice for goodness will not be so easy, it will be made by people on the move in a valley between a clamor of blesses and curses: noise, heat, children, cattle, and someplace else to go, even in the Promised Land. Good and evil are set on the mountains to either side. What will be required will be a radical choice for the good, a choice that is set before them on this very day, *ha Yom*, even though the actual event of choice is yet ahead. But that moment, the key task is to behold, really to see that which is given to you, and that, notes Levinas, is primarily that you are in relationship to a whole attentive, witnessing society.

What one must do is pay attention not only to the commanded laws, but to the collective community whom we see around us—if we look around us—to struggle to understand all at once, to hear our way into a new perception. The horror and evil of slavery are interrupted by the radical liberation into the wide space of freedom, but for a life on the Land, the order that will permit justice must be established. Forty years of wandering, for this moment.

## Seeing the City on the Hill

One might say that this example is too parochial. Let us turn to another proof text, of the one set of English immigrants coming to this land. The Mayflower Compact is a self-conscious rehearsal of the same scene as at River Jordan's banks: pilgrims, farmers, and dreamers just before they entered a new terrain, a land so abundant that John Smith says, "The ground is so fertile that doubtless it is capable of any grain, fruits, or seeds you will sow or plant" (Cheever 1849). Knowing that they are about to move into a promised land, the immigrants were keenly aware of the bloody religious brutality that they had left: thirty years of struggle for this moment[17] (Keenen 1999). As stated in the Mayflower Compact of 1620:

(We) do by these present, solemnly and mutually, in the presence of God and one of another, covenant and combine ourselves together into a civil body politic, for our better ordering and preservation and furthering of the ends aforesaid; and by virtue hereof to enact, constitute and frame such just and equal laws, ordinances, acts constitution, offices from time to time as shall be thought most meet and convenient for the general good of the colony; unto which we promise all due submission and obedience.

Even before immigration, historian Barry Shain reminds us, the first colonists sought to determine communal rules and obligations that would obtain in the new commonwealth. To begin, they set out clear compacts and strong internal relationships. Presented with the emerging model of "individualistic competitive, commercialized, ruthlessly hierarchical social worlds, or a centralized state characteristic of the Renaissance England whence they had fled" (Shain 1994), they looked to the civil body politic, to their faith, and to the faces of "one of another." In other words, they did not leave the new world of total possibilities to chance, but to communities of meaning and discourse to which they were committed and to which they entrusted their best aspirations. I am interested in not only the civil toleration that marked the discourse with distinctive rules that recognized religious difference, but prohibited discrimination relative toward it,[18] but in the idea that new rules would be necessary to deal with exigencies of the physical, tangible possibilities offered by the New World. Beyond toleration lay The City on a Hill, the prophetic notion that upending the world as it was spoken of would allow for new language of human freedom. This idea of the communal, that the sin of each and the obligation of each mattered to the moral fabric, the moral fate of the whole, claims Shain, is as deeply a part of the American proposition as the notions of individualism and autonomy. Morality is both corporate and based in the call of conscience, which is precisely why toleration is so important. And linked to this was the fundamental struggle for justice that we still make in American life: how will great wealth and great freedom from the darkness of the past, and the possibility of this beautiful view that we see be carried by the community? Will immigrants, with troubles and poverty and diseases that will in fact be cured by the land, have to destroy the land to use it? Will immigrant families destroy the lives of the native families? Will the work require the backs and hands of slave families to perform? Can we live on the land in justice? What will the role of women be? It was a question that haunted the early writings of these travelers, and emerges in the writings of religious thinkers. Could the new thing—democracy—also do justice?

These social compacts and this social commonwealth created alliances so strong that rights, and even life, could be forfeited in its defense. It is to these social compacts that we turn, for this is the sort of activity that must undergird the work of truly revolutionary medical research. Here, facing new terrain, new physical landscapes, and new power relationships, what of the past language do we use and what do we abandon? We specifically must ask about the powerless and the powerful, and we must stop and deliberate, despite calls for haste. Can we see the face, need, and cause of one another and the moral seriousness of the arguments that are made for human survival? Will it be possible to place justice in the foreground? After thirty years of war over abortion and all the battle lines we have drawn, could we face radical newness with radical vision and toleration of significant difference so that we could explore the next horizon, aware of the cacophony of calls, and the possibility of sin and of goodness?

Typically and soundly enough, we are moving ahead as best we can with the regulations. The NBAC has called for the establishment of a National Review and Oversight Panel within the U.S. Department of Health and Human Services that would establish a process of evaluating research projects similar to the Recombinant DNA Advisory Committee work with recombinant DNA. This is a critical first step.

But it is not enough. Moral vision has to precede research or we will be constantly in a reactive legal position, seeking to justify what is already unfolding, or struggling to find a political move or linguistic turn to allow political peace. We must have, in addition to a new national commission, a new national conversation as we struggle to create a coherent language to speak of what we do.

It is time to note that the field of bioethics has not been yet bold enough to turn to physicians, in this case researchers, and start by asking moral questions first. I suggest the issues of origins, telos, process, and context must emerge at the outset; ethical questions from our discipline that have to be addressed to scientists.

## Notes for the Next Step

For the purpose of brevity, I will flag only the first two here. First, issues of process are the most familiar ground of our debate, and involve obligations to research subjects whose gametes we intend to use and to the embryos that are used or are created for these purposes. What are the real standards of informed consent, and refusal?[19] But process (O'Neil 1999) is not only the private consent of individual patients, or researchers. Important and temptingly familiar as these questions are, we must develop our obligations to the field by addressing the new problems that arise when ethicists are turned to as experts in the process. How will we be responsible for reporting and sharing data in the altered climate of a health care marketplace shuddering under the impact of managed care and driven by the needs of a enormous pharmaceutical industry? In this work, who will create new rules of engagement, set standards for clinical bioethics created by an open national process?[20] How will moral consultation be understood in a marketplace that is proprietary? Linked to this is a serious struggle with the nature, goal, and meaning of obligations to biotech organizations. What is the proper relationship of ethicists to proprietary companies?[21]

## Justice

Finally, and most centrally important, are our duties and obligations to social justice. We must struggle openly about how we will seriously confront justice issues and, important, what that means for this research; how far is too far? Unless we figure out how to raise the question of justice and what it means, we have not done our work at all. Unless our obligations to society can at all times trump particular needs of a company, we have no business doing business as ethicists.

Two kinds of justice issues must be addressed. The first is procedural—how we will speak to each other in a democratic society where the action of one affect the moral narrative of all. What is the role of consensus in a society that is both pluralistic and often deeply divided over appropriate norms? How can we develop appropriate shared language for public debate and decision making while remaining respectful of differences and accountable to substantive moral disagreements? Roman Catholic moral theologians have insisted that we do not "see" the embryos as members of our human community. But could anything or anyone else be "unseen" by this work?

This last issue is key, not one that can be avoided. It is the substantive issue of social justice. If scientists are correct, and if the entire landscape on which we walk will be new, how do we ensure that this journey remain Exodic? By this I mean, how do we make sure to *see, Re'ah*, that the freed slaves, the widow and the orphan, the stranger, the pilgrim traveler, the refugee, the uninsured, all the witnesses who surround us, have access to this new world? Research done always will mean research foregone. Will this research help or avoid the problem of access to health, given that poverty and poor health are so desperately intertwined in this country? How does the Jewish demand for social justice as an absolute norm affect our obligations? How does the mandate for healing in religious and secular communities unfold? How can difficult issues of global justice and fair distribution be handled in research involving private enterprise? What should societies insist on? Even if we are to be driven by consequences, what should be our consideration for the larger environmental appeal and not just the appeal involving one patient? Since we foreground justice, how will we insist that justice be a part of all the deliberations? What do we mean by this? If a bioethicist were aboard the *Mayflower*, what could she have asked for? Something close to what we have, flawed and shining, I think: a civil body politic, consciousness of each other, a new way to talk, to see, the sense that we are here in the first place by the grace of God, and determination to be a light unto nations, a city on a hill.

### The Banks of Jordan by the Dock of the Bay

A final story: Geron sits in a small building along the western marsh rim of the San Francisco Bay, along with Intel and dozens of biotech companies that genuinely believe in and are committed to a vision of health, working with the magic cells that begin us. Ethicists wait with these hardworking explorers: everything is before us, really, very little has been found, much is still theoretical, it is a beautiful vision. But there is a line of railroad tracks, and just on the other side is the poor, largely African-American and Hispanic community of Belle Haven, and the view of the future from there is far more bleak, less beautiful. If you take a wrong turn you can easily miss the way to Geron (I do it all the time) and get lost in the neighborhood. And perhaps not lost at all, really. It is California poverty, each little broken house needing paint, the health care clinic also needing paint, the many corner churches. There, in the midst of Silicon Valley, live and work the marginalized, excluded, and uninsured. And that question of terrain, geography, must call us again and again to remember who we are, what we do, and for whom we must speak.

# Notes

1. I made the analogy that it is rather like the notion that you have to destroy the village in order to save it. One can certainly hear many arguments that rest on this premise based on the theory of the doomed embryo. For example, in testimony before the NBAC in January 1999, Francoise Bayliss made this point.

2. Technical advances made it possible to enhance the potential for even grades 3 and 4 embryos to implant , we were told, and hence the grading system was not relevant. Many high-grade embryos were being discarded as IVF physicians were trying to implant fewer embryos to achieve pregnancy, hoping to avoid multiple births or fetal-reduction abortions.

3. At this meeting we were also told that mouse embryos were used in the culture medium that supported the newly growing stem cells; yet another issue to think about.

4. Since they were not yet implanted, these early embryos did not have moral status and were regarded halachically as other body tissues, thus existing not only in the liminal state that Jewish and Islamic law takes for the first forty days of conception, but in a state even before that.

5. So is the former, of course, and that problem might also be addressed by this technology.

6. Personal communication, Oakland, CA, 1999.

7. Among them patients with Parkinson's disease and families of those with Alzheimer's disease, spinal cord injury, cancer, and cardiac illness.

8. This is of course, a narrative in which not all of us, nor any of us all of the time, were transfixed by the science. But it was not only the ethicists at Geron; I mean to include colleagues at NBAC, IRBs, and many others who read of the research. It is, in fact, stunningly interesting work.

9. Rather than exotic, which is how we are thinking now.

10. The argument: we know so little about the magic cells, and the potential is so very great. Lives can be saved that are now tragically lost.

11. Or, for example, find the mechanism that allows them to communicate with trophoblastic cells and aggregate, as they do in mice.

12. Language, too, is slippery. If we create new categories and definitions for life and death, can we elude controversy? This mirrors the regeneration with the borders of death in the 1967 Harvard criteria for death by neurologic criteria. When transplantation technology advanced and cadaveric organs could be used safely, we accepted the new terms; in fact, ethicist participated in teaching them. "Brain death" is now widely accepted, even as ethicists question its arbitrariness.

13. Another example is the "use" versus "derivation" language of the NIH, which supports both a conservative view that blastocysts are protected and a progressive view that stem cells are not embryos—a sort of a "don't ask, don't tell" for science.

14. In fact it is our fascination with the consent process that creates other problems.

15. It is not too far afield to note that families do not bring into being every possible gamete. Theoretical children are not born so that the quality of life of other children can be secured.

16. It is a call to attention, and as commentators have noted, it is a call in an odd sort of grammar, a singular call, a call to each, just before the tense changes to the plural ("look, you," changing to "I set before all of you to hear").

17. James Keenan called for us to think ourselves back to a time in the fifteenth century, but is not the sixteenth century the best place to start? My position is based on a feminist understanding that we have to think ourselves back into the seventeenth century, a time that includes women and children and the need to settle a new ground with vulnerable families, not just to claim it.

18. As in the Maryland Doctrine of Toleration, or the Virginia Constitution

19. Such questions are as follows: how do we determine the appropriate locus of consent or refusal when it is a disputed arena? how do we maintain sensitivity to women from whom the tissue taken? is consent ever possible in IVF clinics? what do we mean by "full development of respect" toward the contested entity? who should exercise control over the disposition of fetal or embryo tissue?

20. In the national organization of bioethics, the American Society for Bioethics and Humanities, a broad committee worked to create core competencies for the field.

21. For example, who should constitute an ethics board, and who should serve on it? under what conditions (e.g., remuneration, stock options, etc.)? what is the obligation of the company toward primary commitments: academic freedom, social justice? to whom are we accountable?

## References

Cheever, G. B. 1849. *The Journal of the Pilgrims in New England, in 1620*, 2nd ed. New York.

Geron Ethics Advisory Board. 1999. Research with stem cells: Ethical considerations. *Hastings Center Report* 29: 31–35.

Keenen, J. 2000. Cloning and ethical limits. In: Lauritzen, P., ed. *Beyond Cloning: New Ethical Issues in Reproductive Medicine*. New York: Oxford University Press.

McGee, G., and Caplan, A. L. 1999. What's in the dish? *Hastings Center Report* 29: 36–38.

Nagy, A., Rossant, J., Nagy, R., Abramow-Newerly, W., and Roder, J. C. 1993. Derivation of completely cell culture-derived from early passage embryonic stem cells. *Proceedings of the National Academy of Sciences of the USA* 90: 4824–4828.

O'Neil, O. 1999. Keynote speech at the annual conference of the American Society for Bioethics and Humanities, Philadelphia.

Regalado, A. 1998. The troubled hunt for the ultimate cell. *Technology Review.* MIT Magazine (July, August).

Shain, B. 1994. *The Myth of American Individualism: the Protestant Origins of American Political Thought.* Princeton, NJ: Princeton University Press.

Zoloth, L. 2000. Ethics of the eighth day: Jewish bioethics and stem cell research. In: *National Bioethics Advisory Board Stem Cell Report.* Vol. II. Rockville, MD.

# Glossary

**adult stem (AS) cells**  stem cells found in the adult organism (e.g., bone marrow, skin, intestine) that replenish tissues in which cells often have limited life spans. They are more differentiated than embryonic stem (ES) cells or embryonic germ (EG) cells.

**ART (assisted reproductive technology)**  all treatments or procedures that involve handling human eggs and sperm for the purpose of helping a woman become pregnant. Types of ART include in vitro fertilization, gamete intrafallopian transfer, zygote intrafallopian transfer, embryo cryopreservation, egg or embryo donation, and surrogate birth.

**blastocyst**  a mammalian embryo in the stage of development that follows the morula. It consists of an outer layer of trophoblast to which is attached an inner cell mass.

**blastomere**  one of the cells into which the egg divides after it is fertilized; one of the cells resulting from the division of a fertilized ovum.

**chimera**  an organism composed of two genetically distinct types of cells.

**cloning**  production of a precise genetic copy of a molecule (including DNA), cell, tissue, plant, or animal.

**differentiation**  specialization of characteristics or functions of cell types.

**diploid cell**  cell containing two complete sets of genes derived from the father and mother, respectively; normal chromosome complement of somatic cells (in humans, 46 chromosomes).

**ectoderm**  outer layer of cells in the embryo; origin of skin, pituitary gland, mammary glands, and all parts of the nervous system.

**embryo**  (1) beginning of any organism in the early stages of development, (2) a stage (between ovum and fetus) in the prenatal development of a mammal, (3) in

This glossary is found in the National Bioethics Advisory Commission report, *Ethical Issues in Human Stem Cell Research*, September 1999. The full report is available at www.bioethics.gov.

humans, the stage of development between the second and eighth weeks after fertilization, inclusive.

**embryonic stem (ES) cells**   cells that are derived from the inner cell mass of a blastocyst embryo.

**embryonic germ (EG) cells**   cells that are derived from precursors of germ cells from a fetus.

**endoderm**   innermost of the three primary layers of the embryo; origin of the digestive tract, liver, pancreas, and lining of the lungs.

**ex utero**   outside of the uterus.

**fibroblast**   a cell present in connective tissue, capable of forming collagen fibers.

**gamete**   (1) any germ cell, whether ovum or spermatozoon, (2) a mature male or female reproductive cell.

**gastrulation**   process of transforming the blastula into the gastrula, at which point embryonic germ layers or structures begin to be laid out.

**germ cells**   gametes (ova and sperm) or cells that give rise directly to gametes.

**haploid cell**   a cell with half the number of chromosomes as the somatic diploid cell, such as the ova or sperm. In humans, the haploid cell contains 23 chromosomes.

**in vivo**   in the natural environment (i.e., within the body).

**in vitro**   in an artificial environment, such as a test tube or culture medium.

**in vitro fertilization (IVF)**   a process by which a woman's eggs are extracted and fertilized in the laboratory and transferred after they reach the embryonic stage into the woman's uterus through the cervix. Roughly 70 percent of assisted reproduction attempts involve IVF using fresh embryos developed from a woman's own eggs.

**karyotype**   chromosome characteristics of an individual cell or cell line, usually presented as a systematic array of metaphase chromosomes from a photograph of a single cell nucleus arranged in pairs in descending order of size.

**mesoderm**   middle of the three primary germ layers of the embryo; origin of all connective tissues, all body musculature, blood, cardiovascular and lymphatic systems, most of the urogenital system, and lining of the pericardial, pleural, and peritoneal cavities.

**morula**   (1) mass of blastomeres resulting from early cleavage divisions of the zygote, (2) solid mass of cells resembling a mulberry, resulting from cleavage of an ovum.

**oocyte**   (1) diploid cell that will undergo meiosis (a type of cell division of germ cells) to form an egg, (2) immature ovum.

**ovum**   female reproductive or germ cell.

**pluripotent cells**   cells, present in the early stages of embryo development, that can generate all cell types in a fetus and in the adult and that are capable of self-renewal. Pluripotent cells are not capable of developing into an entire organism.

**preimplantation embryo**   (1) embryo before it has implanted in the uterus, (2) commonly used to refer to in vitro fertilized embryos before they are transferred to a woman's uterus.

**somatic cells**   [from *soma*, the body] (1) cells of the body which in mammals and flowering plants normally are made up of two sets of chromosomes, one derived from each parent, (2) all cells of an organism with the exception of germ cells.

**stem cells**   cells that have the ability to divide indefinitely and to give rise to specialized cells as well as to new stem cells with identical potential.

**totipotent**   having unlimited capacity. Totipotent cells have the capacity to differentiate into the embryo and into extraembryonic membranes and tissues. Totipotent cells contribute to every cell type of the adult organism.

**trophoblast**   outermost layer of the developing blastocyst of a mammal. It differentiates into two layers, the cytotrophoblast and syntrophoblast, the latter coming into intimate relationship with uterine endometrium with which it establishes a nutrient relationship.

**zygote**   (1) cell resulting from fusion of two gametes in sexual reproduction, (2) fertilized egg (ovum), (3) diploid cell resulting from union of sperm and ovum, (4) developing organism during the first week after fertilization.

# Contributors

**Françoise Baylis** is associate professor of medicine and philosophy at Dalhousie University, Halifax, Canada. She is currently a member of the Canadian Biotechnology Advisory Committee, an executive member of the National Council on Ethics in Human Research, and a member of the Biomedical Ethics Committee of the Royal College of Physicians and Surgeons of Canada. Her areas of research interest include the ethics of new genetic technologies, research involving humans, and women's health.

**Cynthia B. Cohen**, Ph.D., J.D., is senior research fellow at the Kennedy Institute of Ethics at Georgetown University, Washington, D.C., and adjunct associate at the Hastings Center in Garrison, New York. She was formerly executive director of the National Advisory Board on Ethics in Reproduction and chair of the philosophy department at the University of Denver. She has also chaired several bioethics committees for the Episcopal Church. Her publications include *New Ways of Making Babies* and *Wrestling with the Future: Our Genes and Our Choices*, as well as numerous articles on ethical issues that arise at the beginning and end of life.

**Rabbi Elliot N. Dorff**, Ph.D., is Rector and Distinguished Professor of Philosophy at the University of Judaism in Los Angeles. He is Vice-Chair of the Committee on Jewish Law and Standards of the Conservative Movement in Judaism. His books include *Matters of Life and Death: A Jewish Approach to Modern Medical Ethics* (Philadelphia: Jewish Publication Society, 1998).

**Margaret A. Farley** is Gilbert L. Stark professor of Christian ethics at Yale University Divinity School. She is the author or editor of five books and more than seventy articles in the areas of medical ethics, sexual ethics, historical Christian ethics, contemporary Roman Catholic ethics, and feminist ethics.

**John C. Fletcher** is Professor Emeritus of Biomedical Ethics and Internal Medicine at the University of Virginia School of Medicine. From 1987 to 1997, he directed the School of Medicine's Center for Biomedical Ethics. From 1977 to 1987, he was Chief of the Bioethics Program at the Warren G. Magnuson Clinical Center of the National Institutes of Health.

**Suzanne Holland** is assistant professor of religious and social ethics at the University of Puget Sound, Tacoma, Washington. Holland is chair of the Religion and Health Care Group of the American Academy of Religion, an assistant bioethics editor for *Religious Studies Review*, and co-chair of the 2000–2001 Program Committee of the American Society of Bioethics and Humanities. Her research interests range from the ethics of human genetics, biotechnology, and Internet technology, to broader issues in religion, culture, and gender.

**Karen Lebacqz**   is Robert Gordon Sproul professor of theological ethics at Pacific School of Religion and the Graduate Theological Union, Berkeley, California. She is a former member of the National Commission for the Protection of Human Subjects of Biomedical and Behavioral Research, and is chair of the Ethics Advisory Board of Geron Corporation, Menlo Park, California. Author of more than six books and numerous articles, she specializes in bioethics, professional ethics, and theories of justice. She is also an ordained minister in the United Church of Christ.

**Glenn McGee**   is an associate director at the Center for Bioethics of the University of Pennsylvania, where he holds appointments in Philosophy, History and Sociology of Science, and Cellular Engineering. He is editor-in-chief of the *American Journal of Bioethics* and author of a number of articles about ethical issues in genetics, stem cell research, and reproduction. His books include *The Perfect Baby* (1997), *The Human Cloning Debate* (1998), and *Pragmatic Bioethics* (1999). He is currently writing a book about the impact of life-extension technologies on concepts of human nature in religion and social life.

**Margaret R. McLean, Ph.D.,**   is director of biotechnology and healthcare ethics at the Markkula Center for Applied Ethics at Santa Clara University, and director of the Applied Ethics Center at O'Connor Hospital in San Jose. She also teaches ethics in the Religious Studies Department of Santa Clara University. She is a member of the Bioethics Committee at O'Connor Hospital, the Infant Bioethics and Infant Care Review Committees at Santa Clara Valley Medical Center. She is on the editorial board of the *Religious Studies Review*. Current projects include participation on the California State Advisory Committee on Human Cloning and a manuscript on genetic selection.

**Gilbert Meilaender**   has taught religious ethics at the University of Virginia and at Oberlin College, where he held the Davis chair in religion. He is currently Duesenberg professor of christian ethics at Valparaiso University. Among his publications is *Body, Soul, and Bioethics* (1995).

**Michael M. Mendiola, Ph.D.,**   is associate professor of Christian ethics, Pacific School of Religion in the Graduate Theological Union, Berkeley, California. He is a member of the Geron Ethics Advisory Board and convener of the Bay Area Faith and Health Consortium, an interdisciplinary program exploring conceptual and practical interrelationships among faith, spirituality, and human health. He has published in the area of bioethics.

**Thomas B. Okarma, Ph.D., M.D.,**   is Geron Corporation president and chief executive officer. Before joining Geron, he was an assistant professor of medicine at Stanford University School of Medicine and subsequently founder and chairman-CEO of Applied Immune Sciences (AIS), a cell and gene therapy company. AIS was acquired by Rhone-Poulenc Rorer (RPR) in 1995, at which time Dr. Okarma was vice president at RPR until joining Geron Corporation in 1997.

**Erik Parens** is the associate for philosophical studies at the Hastings Center, a bioethics think tank in Garrison, New York. He has published extensively on the ethical and social questions raised by biotechnologic advances; he is also editor of *Enhancing Human Traits: Ethical and Social Ramifications* (1998) and co-editor of *Prenatal Genetic Testing and the Disability Rights Critique* (2000).

**Ted Peters** is professor of systematic theology at Pacific Lutheran Theological Seminary and the Graduate Theological Union, Berkeley, California, and a research associate at the Center for Theology and the Natural Sciences. He is editor of *Dialog, A Journal of Theology*. He is author of *GOD—The World's Future*, 2nd ed. (2000); *Playing God? Genetic Determinism and Human Freedom* (1997); and *For the Love of Children: Genetic Technology and the Future of the Family* (1996). He is editor of and contributor to *Genetics: Issues of Social Justice* (1998) and *Science and Theology: The New Consonance* (1998).

**Thomas A. Shannon** is professor of religion and social ethics in the Department of Humanities and Arts at Worcester Polytechnic Institute, Worcester, Massachusetts. He is the author of several books and articles in the area of bioethics.

**James A. Thomson** is a University of Wisconsin–Madison developmental biologist in the Department of Anatomy in the School of Medicine. He also serves as the chief pathologist at the Wisconsin Regional Primate Research Center on the Madison campus. Since joining the Wisconsin Regional Primate Research Center, he has conducted pioneering work in the isolation and culture of nonhuman primate and human embryonic stem cells, undifferentiated cells that have the ability to become any of the cells that make up the tissues of the body. Dr. Thomson directed the group that reported the first isolation of embryonic stem cell lines from a nonhuman primate in 1995, work that led his group to the first successful isolation of human embryonic stem cell lines in 1998.

**Paul Root Wolpe,** Ph.D., a medical sociologist, is a fellow of the Center of Bioethics of the University of Pennsylvania, where he holds faculty appointments in the Departments of Psychiatry and Sociology. He is also a senior fellow of Penn's Leonard Davis Institute of Health Economics. He is the author of *Sexuality and Gender in Society* and the forthcoming *Manual for End-of-life Decision-Making*.

**Ernlé W. D. Young** is professor of medicine (ethics) in the Stanford University School of Medicine, where he has taught for twenty-seven years. He is also codirector of the Stanford University Center for Biomedical Ethics, which he cofounded in 1989. He has published widely and lectured nationally and internationally on ethical issues in medicine and the life sciences.

**Laurie Zoloth** is professor of social ethics and Jewish philosophy, and director of the Program in Jewish Studies at San Francisco State University. Her research is focus on bioethics, and in 2000–2001 she is president of the American Society for Bioethics and Humanities. Her books include *Health Care and The Ethics of Encounter: A Jewish Discussion of Justice; Notes from a Narrow Ridge: Bioethics and Religion,* coedited with Dena Davis; and *The Margin of Error: The Inevitability, Necessity and Ethics of Mistakes in Medicine and Bioethics,* coedited with Susan Rubin.

# Index